Kritische Verbraucherforschung

Reihe herausgegeben von
C. Fridrich, Wien, Österreich
R. Hübner, Klagenfurt, Österreich
K. Kollmann, Wien, Österreich
M.-B. Piorkowsky, Bonn, Deutschland
N. Tröger, Wien, Österreich

In der Reihe „Kritische Verbraucherforschung“ werden Sammelbände und Monographien veröffentlicht, die im Gegensatz zur herkömmlichen, am Markt-Kauf-Paradigma orientierten Verbraucherforschung wenig reflektierte Ansätze stärker in den Blick nehmen. Kritisch wird vor allem die tendenziell zunehmende Vereinnahmung sämtlicher Lebensbereiche durch den Markt gesehen. Die zunehmende Marktorientierung verändert unsere Gesellschaft insofern massiv, als sie erstens Menschen einseitig marktabhängig macht, zweitens aufgrund des dem herrschenden ökonomischen Verständnis inhärenten Wachstumsparadigmas dazu führt, dass alle Lebensbereiche nach der Wachstumslogik funktionieren (müssen) und dadurch drittens die Zerstörung unserer Lebensgrundlagen zu beschleunigen scheint.

Kritische Verbraucherforschung wird daher in mehrerlei Hinsicht aus einer emanzipatorischen Haltung heraus entwickelt: wider die Instrumentalisierung der Verbraucherforschung, wider die Vermarktlichung der Gesellschaft, wider die Infantilisierung der Verbraucherinnen und Verbraucher und wider die Privatisierung von politischer Verantwortung.

Mit einer Kritischen Verbraucherforschung eng verknüpft ist eine kritische Verbrauchertheorie, in der jene Kontexte angemessen berücksichtigt werden, in die das Handeln von Verbraucherinnen und Verbrauchern eingebettet ist, nämlich private Haushalte, Erwerbstätigkeit und Gesellschaft. Diese Handlungskontexte sind für eine sozioökonomisch ausgerichtete Wirtschafts- und Verbraucherbildung von höchster Relevanz, in der die gesellschaftlich eingebetteten Akteurinnen und Akteure in den Mittelpunkt der Analyse gerückt werden und die auf kompetente Orientierungs-, Urteils- sowie Handlungsfähigkeit abzielt.

Die Reihe ist auch ein Angebot an die Verbraucherbildung, die Verbraucherpolitik und die Nachhaltigkeitspolitik, insofern diese Bereiche auch Interesse an wissenschaftlichen Erkenntnissen und Ansätzen abseits von rein marktökonomischen Perspektiven und Gestaltungsspielräumen haben. Umgekehrt sind auch Vertreterinnen und Vertreter insbesondere der Bildungs-, Verbraucher- und Nachhaltigkeitspolitik eingeladen, mitzuwirken und mit ihren Fragestellungen wiederum die Kritische Verbraucherforschung zu fordern und zu fördern.

Weitere Bände in der Reihe http://www.springer.com/series/13869

Isabel Cruz · Rafaela Ganga
Stefan Wahlen
Editors

Contemporary Collaborative Consumption

Trust and Reciprocity Revisited

Springer VS

Editors
Isabel Cruz
Matosinhos, Portugal

Rafaela Ganga
University of Liverpool
Liverpool, UK

Stefan Wahlen
Wageningen University
Wageningen, The Netherlands

Kritische Verbraucherforschung
ISBN 978-3-658-21345-9 ISBN 978-3-658-21346-6 (eBook)
https://doi.org/10.1007/978-3-658-21346-6

Library of Congress Control Number: 2018944365

Springer VS

Printed on acid-free paper

This Springer VS imprint is published by the registered company Springer Fachmedien Wiesbaden GmbH part of Springer Nature
The registered company address is: Abraham-Lincoln-Str. 46, 65189 Wiesbaden, Germany

Contents

Preface: Contemporary Collaborative Consumption. Trust and Reciprocity Revisited

With increasing digitalization of everyday life, consumers' access to goods and services in the online world is expanding. Not only do these developments influence how people interact and socialize, but the digitalization also extends more profoundly to how we organize daily lives, identities and sustenance. One of the overarching terms describing this change is the sharing economy. With regard to consumer activity in the sharing economy it is possible to speak of collaborative consumption. The new forms of consuming described as collaborative consumption are assumed to mitigate environmental and social outcomes of consumer capitalism and accordingly positively contributing to more sustainable development. A conflicting perspective portrays collaborative consumption in exacerbating new forms of inequalities manifesting in a form of turbocharged capitalism with individualized and tailored on-demand production and consumption. Looking at the empirical reality within these two extreme positions, the book series on "Critical Consumer Studies" is a well-fitting outlet illustrating cases of collaborative consumption in a European context.

The series "Critical Consumer Studies" is interested in ways of theorizing consumers and consumption beyond the prevailing paradigm focusing on markets and acts of purchasing. Indeed, the book series criticizes the centrality of a buyer's role and individual behavior. Consequently, "Critical Consumer Studies" is interested in scrutinizing consumption in situated everyday contexts and in considering implications for consumer education and consumer policy. This volume addresses the critique of the one-dimensional consumer image by examining societal developments coined as sharing economy and seemingly novel forms of collaborative consumption.

The distinction of economic categories as consumers and producers are vanishing in times of increasing digitalization and identification of individuals as both, producer and consumer. Being considered a solution to slow down processes of destroying our livelihood, but also as subordinating areas of life to an efficiency and growth paradigm, the empirical reality of collaborative consumption is more diverse. This volume provides empirical cases of collaborative consumption oscillating around the concepts of trust and reciprocity. The authors outline aspects of collaborative consumption away from the prevailing market paradigm and contribute to critical consumer studies by explicating the role of trust and reciprocity in exchanges of increasingly digitalized everyday life.

Isabel Cruz, Rafaela Ganga and Stefan Wahlen

Porto, Liverpool and Wageningen, February 2018

Vorwort: Zeitgenössischer kollaborativer Konsum. Vertrauen und Reziprozität überdenken

Mit einer zunehmenden Digitalisierung des Alltags expandiert auch der Zugang von Verbrauchern zu Gütern und Dienstleistungen über die digitale Welt. Diese Entwicklungen beeinflussen nicht nur wie Verbraucher interagieren und Kontakte knüpfen, sondern die Digitalisierung erstreckt sich auch auf weite Teile der Organisation des täglichen Lebens und manifestiert demgemäss Identitäten und der Existenz. Einer diese Entwicklungen umschreibende Begriffe ist der der Sharing Economy. Im Hinblick auf Verbraucheraktivitäten in der Sharing Economy wird oft von kollaborativem Konsum gesprochen werden. Auf der einen Seite wird angenommen, dass kollaborativer Konsum ökologische und soziale Konsequenzen des Konsumkapitalismus abschwächt und entsprechend zu einer nachhaltigeren Entwicklung beitragen kann. Auf der anderen Seite arbeitet eine widersprüchliche Perspektive heraus, dass kollaborativer Konsum Formen von Ungleichheiten verschärft und als eine Form des turbogeladenen Kapitalismus daherkommt welcher individuelle und maßgeschneiderte Produktion und Konsumtion auf Abruf propagiert. Die Buchreihe „Kritische Verbraucherforschung" ist ein passendes Forum um sich mit der empirischen Realität dieser beiden extremen Positionen zu beschäftigen. Dieser Sammelband präsentiert Fallstudien des kollaborativem Konsums in einem europäischen Kontext.

Die Reihe „Kritische Verbraucherforschung" interessiert sich für die Konsumtheorie jenseits des vorherrschenden Markt-Kauf-Paradigmas, das sich auf Märkte und Kaufkte konzentriert. Durchaus kritisiert die Buchreihe die Rolle des Verbrauchers als Käufers. Folglich ist eine „Kritische Verbraucherforschung" daran interessiert, den Konsum in situierten Alltagskontexten zu untersuchen und Konsequenzen für die Verbraucherbildung und die Verbraucherpolitik zu berücksichtigen. Dieser Band befasst sich mit der Kritik des eindimensionalen Verbraucherbildes, indem er zeitgenössische gesellschaftliche Entwicklungen untersucht. Die mit zunehmender Digitalisierung einhergehende Sharing Economy und scheinbar neuartige Formen kollaborativen Konsums sind dementsprechend von höchster Relevanz für eine „Kritische Verbraucherforschung".

Die Unterscheidung von ökonomischen Kategorien der Konsumenten und Produzenten verschwimmen in Zeiten zunehmender Digitalisierung und der Sharing Economy. Die in diesem Bande herausgearbeitete empirische Realität des kollaborativen Konsums zeichnet ein Bild wie die zunehmende Digitalisierung des Alltags einen Beitrag zur Verlangsamung von lebenszerstörenden Prozessen leistet, aber auch Lebensbereiche einem Effizienz- und Wachstumsparadigma als unterwirft. Dieser Band bietet empirische Fallstudien kollaborativen Konsums, welche um die theoretischen Konzepte von Vertrauen und Reziprozität drehen. Die Autoren skizzieren Aspekte des kollaborativen Konsums abseits des vorherrschenden Markt-Kauf-Paradigmas und tragen zu kritischen Verbraucherforschung bei indem sie die Rolle von Vertrauen und Gegenseitigkeit im Austausch des zunehmend digitalisierten Alltagslebens erklären.

Isabel Cruz, Rafaela Ganga und Stefan Wahlen

Porto, Liverpool und Wageningen, Februar 2018

Contemporary collaborative consumption: An introduction

Isabel Cruz, Rafaela Ganga and Stefan Wahlen

The recent decade has seen unprecedented societal challenges. Since the beginning of the economic and social crisis in the global North a decade ago, consumers have started rethinking their everyday lives. Consumers have not only become more mindful in their spending habits, but also more ingenious in living on a limited budget. In times of crisis, individuals have to an increasing extent started thinking about their way of life and the social, economic and natural environment in which they live. The changes that took place during and after this crisis appear to have endured. Individuals and groups of consumers started rethinking ownership and the possibility of accessing rather than owning particular goods and commodities. Moreover, service provision has increasingly attracted modes of sharing rather than provisioning via a traditional market setting.

A term that has been rediscovered to coin these developments in provisioning goods and services is collaborative consumption. This renaissance of alternatives and ancient systems of provision in the broad capitalist market economies raises questions regarding the quality of life and what the future holds for both individuals and communities. The economic downturn in Europe has given rise to increased debate on what future world is envisioned. The United Nations responded with the development of, and agreement upon, Sustainable Development Goals to deal with global injustice and environmental burdens. The sustainable development goals that have been agreed by world leaders in 2015 envision a society without poverty, with a lower impact on ecosystems and a liveable environment for all. Achieving sustainable development by the year 2030 appears imperative.

In that sense, the new forms of sharing that emerged during the crisis, particularly collaborative consumption, are expected to provide possible solutions on the way to more sustainable development. On the other hand, since collaborative consumption is associated with the sharing economy (e.g. Wahlen and Laamanen 2017), critics also stress the challenge of a turbo capitalist approach, bearing in

I. Cruz et al. (Hrsg.), *Contemporary Collaborative Consumption*,
Kritische Verbraucherforschung, https://doi.org/10.1007/978-3-658-21346-6_1

mind new forms of labour injustices and workers' conditions deteriorating. Hence different versions of the sharing economy and collaborative forms of consumption are discussed, one pertaining to a more affirmative and progressive version, seeing solutions to world problems in the promotion of social cohesion and alleviating the environmental burden through collaborative consumption. In contrast, there is a more sceptical and daunting version, which sees collaborative consumption as an exploitative practice that exacerbates existing inequalities and injustices, as well as promoting individual self-interest. Yet social reality is far more complex than these polarized positions. There are ample examples of commercially oriented collaborative consumption which nevertheless impacts the social sphere, demonstrating that the variety of forms of collaborative consumption changes the way that individuals consume collaboratively and thus helps to mitigate the societal challenges present in contemporary society.

Alongside social, economic and environmental developments in recent decades, there has also been a socio-technological shift, with increased digitalisation having an impact on the ways individuals organise their everyday lives and consumption trends. Socio-technological changes associated with the Internet allow for new ways of communicating and interacting in the social sphere, which has impacted on the everyday lives of individuals and communities. Since the 1990s, the Internet has gained ground, and new ways of communicating have generated new forms of interaction. Together with e-mails, social media platforms such as Facebook, Twitter, and LinkedIn have become more popular, with individuals becoming increasingly interconnected. Using platforms to stay in touch with friends or to share a particular identity or lifestyle – e.g. sharing music – allows individuals to express themselves not only in the analogue world, but to an increasing extent in a digital version of the self. Digitalisation has also brought, equally, new forms of exchange and market organisation, allowing consumers to interact and exchange goods and services far more widely than in analogue times.

Digitalisation has also affected working life: the boundaries of leisure and working time are blurring: all-day access to e-mails from anywhere in the world, or remote forms of work are just two examples. This fact is also inseparable from the increasing use of portable electronic devices. The rise of smartphones, tablets and e-readers has been a major factor in decreased computer use as the primary means of accessing the Internet. Over the past decade, the use of smartphones and tablets has changed the role of the Internet in daily life, enabling us to communicate with each other from anywhere, buy the latest fashion, order a taxi, do grocery shopping, show friends what we are eating, etc. and stay continuously connected. This list could be broadened to an unlimited array of everyday activities that have changed as digitalisation has taken hold of contemporary life. With the rise of new kinds of mobile devices since the early 2000s, the use of mobile phones and apps has changed the social grid and the way we acquire commodities and provide services. Not only do we exchange information, or buy and sell commodities. These new technological devices also enable consumers to manage their everyday life in new ways: it is not only possible to shop for groceries or other items, but also to access a variety of other services such as banking and paying tax, for which an in-person visit is no longer required. These things can be done from the couch in your living room.

Contemporary collaborative consumption can accordingly be understood as a relational practice, since individuals consume goods and services while at the same time relating to each other (e.g. meals, travelling, working, school). These 'practices of collaborative consumption' must be analysed in relation to the temporal and geographical context as well as the social and natural environments in which they occur (Felson and Spaeth, 1978). In contemporary societies of the global North, consumers find more satisfaction and status in experiencing, rather than owning material goods. In digitalised societies, collaborative forms of consumption are often characterized and mediated by digital networks and technologies. In the digital age, individuals engage in collaborative forms of consuming in order to pursue their self-interest while also considering the collective good, eventually making a move towards more sustainable consumption on a larger scale.

This book is interested in the social, environmental and (to a lesser extent) economic aspects of contemporary forms of collaborative consumption. The collection of chapters in this book comprises a selection of European case studies. The book discusses the role of collaborative consumption and the new sharing ethos in contemporary society. The book addresses the way alternative and new forms of collaborative consumption are socially shared and understood and how these unfold and change existing thinking about consumption and production as well as work and leisure. The authors are not only interested in innovative business ideas, but in the attitudes of consumers as well as the societal implications of a sharing ethos. We address possible implications and interdependencies for consumer landscapes and the sharing economy in a future society. In this way, we seek to engage in critical debates to promote change towards a more sustainable future, one which acknowledges a diverse set of ideas, approaches and principles in a broad array of systems of provision.

In the aftermath of the multiple crisis outlined above and an increasingly digitalised everyday life, contemporary collaborative consumption is propelling a contemporary wave towards alternative modes of everyday activity but also advancing alternative understandings of consumption. In the times of crisis we have witnessed over the last decade, bartering has become a more common mode of exchange in our societies. Collaborative consumption is a way of accommodating needs and desires in a more sustainable way and possibly lightening the burden for individuals, communities and the wider environment (Botsman and Rogers 2010). Collaborative consumption may bring benefits for the parties involved, in different fields: in economic terms (savings), in social cohesion (relationships or friendships), and in terms of its environmental impact. Collaborative consumption allows individuals to make money from dormant resources in underutilized goods and equipment. In collaborative consumption, the sharing of goods and services may reduce usage costs in economic and environmental terms. Sharing is the basis of collaborative consumption in the sharing economy. Sharing means getting in touch with other people and not just extracting resources from the market system in the form of commodities as suggested by traditional economic and capitalist thought.

Collaborative consumption is in fact shaking up seemingly well-established economic categories, in various ways. The developments outlined above appear to challenge traditional (economic) conceptions which see consumption as the market exchange of commodities. The newly generated interest in alternative forms of exchange challenge the concept of ownership and involve new modes of using and accessing goods (Bardhi and Eckhardt 2012). Collaborative consumption needs to be seen in wider social, economic and environmental contexts. Consumers are showing a robust appetite for collaborative consumption in a sharing-based economy. Collaborative consumption revives ancient usages such as barter and swaps, coping with situations of crisis and translating these concepts for the digital age. As also outlined above, the traditional economic dichotomy of consumers and producers is being eroded as the field of work extends to what in earlier times was regarded as leisure time.

In traditional economic understanding, workers produce and consumers consume. This dichotomist understanding, dividing the economy into consumption and production, is questionable (Piorkowsky 2017; Fridrich et al. 2014). In collaborative consumption, the concept of the 'prosumer' (Toffler 1980) has been revived as the economic categories of producer and consumer have started to merge in collaborative forms of consumption in the sharing economy. This conceptual amalgamation opens up new possibilities on the road to sustainable development. Traditional roles in the sphere of consumption assign roles to consumers and producers. In collaborative consumption, resources may be used in more efficient ways, at least to some extent. It is precisely these dormant resources which are assumed to solve some of the problems raised by the UN's sustainable development goals. Even though certain modes of contemporary collaborative consumption encourage development towards those goals, there are also counter-trends which have 'rebound effects' in the form of greater environmental burdens and antagonisms rather than solidarity and social justice.

In enabling more social ways of consuming, contemporary collaborative consumption reduces alienation. At the same time, collaborative consumption still operates within capitalist market economies and to some extent also contributes to new forms of alienation, for instance in terms of work. Consumer activity in the sharing economy might be interpreted as entrepreneurial activity. To the extent that collaborative consumption takes place in peer-to-peer networks, consumers are taking over the role of the providing market participant. In this sense, collaborative consumption is assumed to involve new business models, e. g. where individuals rent out an apartment or share a ride and thus move into what is conventionally regarded as the producer side of the economic system (Piorkowsky 2017). The emergence of new business models such as co-working, crowdfunding, crowdlending or crowdequity may challenge traditional economic models and their associated representations of how (labour) markets are organized (Laamanen, Barros and Islam 2018).

Stakeholders in business ventures have shown increasing interest in these new models as ways of generating new, large-scale profits as the number of individuals participating in the sharing economy has grown. However, there are also downsides to the new ways of organizing collaborative forms of consumption: the contemporary sharing economy is marked by increasing inequalities, for example between those who give and those who receive, leading to unfairness when providers of a particular service lose their job security. A prominent example of this is the replacement of taxi drivers by ride-sharing services. The micro-entrepreneurial activity in ride-sharing is not sharing in the colloquial sense, but represents a new form of income which threatens unionised markets such as those of the taxi companies. Those who work for those companies see their rights eroded, because they are now seen as (micro-)entrepreneurs.

From an environmental perspective, there may be both advantages and disadvantages to collaborative consumption. It is widely believed it may lead to more intense resource usage as the number of users for a product or service grows. The car or drill is not used for just a few minutes – so in collaborative consumption total usage time increases. Contrarily, even though an empty apartment being rented out to tourists while the owner is away leads to increased usage, a new problem arouses with entrepreneurs reducing the amount of accommodation available to local residents, as has occurred in major European cities. Collaborative consumption does not therefore always lead to a decrease in levels of consumption, since the resulting savings are often used in the acquisition of other goods and services (Vaquero and Calle 2013), possibly leading to a rebound effect. Hence, the ideal picture painted by proponents of collaborative consumption is of a horizontal structure in which there would be no power-induced hierarchy and materials such as space or tools would be shared, as well as immaterial resources such as time (Arcadiano 2018). It is possible to think of shared production and distribution with dispersed people and organizations in collaborative consumption. In this vein, collaborative consumption is seen to be more efficient in the allocation and life-time usage of resources by means of redistribution. New sharing initiatives such as repair cafés promote the recycling, repair and re-use of goods. Collaborative consumption may have opposite effects: it has not yet been proven that these alternative forms of sharing activity are different from traditional forms of capitalism and consumption. Accordingly, we need to critically assess the activities of collaborative consumption as there are tensions in the approaches, institutions and ways of organizing. This book seeks to fill some of the gaps in knowledge of the interplay of these opposing forces and of the many nuances in collaborative consumption.

The central aim of this book is to shed empirical and theoretical light on the many contemporary forms of collaborative consumption. We are interested in the nuances of social and economic activity across and beyond traditional markets, the interplay of the various actors involved in collaborative consumption and the propelling of social change towards more sustainable consumption. We are also interested in establishing to what extent crisis-derived collaborative consumption practices focused on sharing and bartering challenge prevailing productionist economic models and their associated growth paradigm. Based on this overarching objective, other relevant questions arise: in what ways has collaborative consumption impacted society, the environment and the economy? How can contemporary concepts of sharing and collaborating extend existing ways of living and consumer agency? What are the practices, perceptions, and motivations of those involved in collaborative consumption in the current societal context? Which interdependencies of individual and collective action in collaborative consumption and a sharing economy encourage more sustainable development?

In contemporary societies marked by singularities and individualism, collaborative consumption may help consumers and citizens to do more with less and to promote social ties. Increasingly, the concepts of 'do it yourself' and 'change from the bottom up' promoted by certain social or lifestyle movements with a view to enhancing collaborative practices have gained ground in discussion of the responsibility for consumer activity and the role of the state and political parties. In this book, we assume that the new digital technologies promote individual consumers' contribution to the public good. Individual consumers look after collective interests in the sense that social networks encourage the sharing of goods and services among strangers. Collaborative consumption goes viral by placing different types of actors and forms of exchange in relation to each other, and sharing concern with both the environment and sustainable development. In this way, collaborative consumption is to some extent moving away from the monetarized and capitalist market economy approach to consumption. We have organized the chapters of this book in line with the concepts of trust and reciprocity in the sharing economy, with the aim of moving beyond conventional business models and providing a broader social science perspective on collaborative consumption. The concepts of trust and reciprocity as discussed in sociological debates lay the foundations for this, but also set collaborative consumption in a broader social scientific context. The role of trust and reciprocity in social relationships is discussed from various angles in the different chapters of this book. These concepts allow us to critically assess collaborative consumption and its role in societal developments. Each of these concepts enables us to look at how contemporary collaborative consumption practices, enhanced by new technologies and social networks, have altered traditional forms of capitalist market exchange and sharing.

Trust in collaborating and sharing

In the traditional understanding of a capitalist market economy, consumers demand products from the market (Fridrich et al. 2014). Market exchange is made possible with money, which facilitates the exchange of goods and services and transforms them into commodities, thereby alienating producers and consumers. In terms of vital social relations, money creates a distance between producer and consumer. Yet, money also plays a vital role in avoiding risks and uncertainties of exchange both in markets and beyond. Some collaborative consumption initiatives provide alternative repositories of value, such as time used as the medium of exchange in time banking (Arcidiano 2018; Laamanen, Wahlen and Campana 2015). Indeed, other forms of exchange, not involving money and traditional markets, have emerged in collaborative consumption. These new forms of exchange are based on different mechanisms of trust, as the example of timebanking elucidates (Arcidiano 2018). Time as valuation might mitigate inequalities in access to goods and services by enabling consumers to contribute their expertise within a social network. The mutuality of the exchange is mediated by a platform which creates trust as a repository of value.

In contemporary collaborative consumption, strangers exchange goods and services without having to rely on money as a medium of exchange. These exchanges are often intermediated through (virtual) social networks or platforms, which make it easier for strangers to communicate and exchange with each other. Individuals can establish a virtual identity, generating a feeling they actually know each other and reducing risk and uncertainty in exchanges, hence allowing for trust in the unknown other. Bonds created through the act of sharing property or personal belongings require trust. In collaborative consumption, exchange practices have to be accommodated in a relationship of trust between the different parties involved. Saving money and sparing the environment by sharing goods and services also makes it more attractive to trust strangers. There are several mechanisms for building trust beyond monetary exchange: systems of verification, feedback and evaluation on the platform and in networks, but also collaboration with insurance companies. In this book, we seek to explore the social aspects of trust in contemporary collaborative consumption.

Trust is acknowledged as a crucial element for building social relationships. Trust relies on the integrity of a person. But how do we get to trust strangers to share our car/house or exchange goods and services with them in collaborative consumption? The digital social network's ability to create social cohesion and to influence the social process is a major contribution of collaborative consumption. Establishing trust with strangers will endow the relationship with greater importance, and participants in collaborative consumption are willing to sacrifice and to commit to it. With regard to contemporary forms of trust in collaborative consumption, the key factor is the positive or negative consequences of behaving or not according to a particular desired scheme of action. The decision to reciprocate, i. e. to keep trusting, depends on the other party's interpretation of one's behaviour.

Social history can play a significant role in reinforcing individuals' predisposition to trust. Knowing how trust has been sustained within an organization in the sharing economy may influence economic activity and the relationship between the parties involved. In the first instance, exchange processes offer the possibility of trusting or not trusting the other party. The other party can then decide to trust and invest in order to achieve mutual gains. At this point both parties are put at risk, even though in the final analysis they both benefit through reciprocity rather than not investing in and trusting each other (Berg et al. 1994). Reciprocal behaviour can benefit the community, and that is also a reason for greater reciprocity between the parties, even without personal gain. In their contribution on trust, Mortara and Roberti (2018) outline how young people involved in collaborative consumption distinguish between particularized and generalized trust, emphasizing that a network of repeated exchange relationships is a source of trust. They focus on how users protect themselves from the risks associated with the content and extent of the interaction with strangers with whom they have never had to deal before. Trust mechanisms are common on collaborative consumption platforms in order to minimize the risk of sharing among strangers and to encourage people's engagement with collaborative consumption.

In their chapter, Settifi and Lazzer (2018) outline the role of trust in car sharing. Their case study on 'blablacar' sees trust as an essential component in transport platforms. Trust is regarded as a feeling that depends on users' experiences and changes over time and, although the technological infrastructure of car sharing platforms must be trustworthy, there are other factors to be considered in the analysis of the sharing ethos. In a similar vein, Wahlen (2018) considers trust in food sharing communities by reflecting on the notion of individualized collective action. His contribution highlights the role of trust in the community to provide for proper food, but also trust in the public authorities with regard to food safety. Collaborative consumption communities complement trust measures which are individual and also collective. Organisational learning is a tool to provide for food in dispensing locations in order to consume food that is still edible. This assists in reflecting on how lifestyles aiming to move away from traditional markets also build on trust mechanisms beyond money.

Paltrinieri and Spillare (2018) build on the idea of reciprocal trust in the emerging New Food Economy. Their main concern is to deepen knowledge of how trust-building processes may affect collaborative consumer agency. Their analysis focuses on key elements of the trust-building process within Italian Civic Food Networks. They emphasize the ability to create and maintain "active trust" (Giddens 1994) as the core process of resilience and empowerment in local communities. To sum up, the decision to reciprocate is constrained by trust, social history, kindness and the perceived intentions of the other party. This creates interdependency between trust and personal and environmental gains in contemporary collaborative consumption. The aim of this book is to clarify how trust can be combined with, and mutually constrains, consumers' activity in the context of collaborative consumption and the sharing economy over time, and furthermore how trust is considered in terms of reciprocity, the second overarching theme in this book.

Reciprocity in collaborative consumption and production

In contemporary western societies, monetary systems influence how goods and services are exchanged, in contrast to ancient times, when barter predominated. Money takes on a central role in understanding reciprocity as the expectation of getting something in return. Yet there is more to exchange than reciprocating by means of money, just as there is more to reciprocity than simple exchange. In advanced capitalist societies reciprocity is embedded in complex systems of exchange. Indeed, in contemporary societies, reciprocity may help to highlight the social aspects of exchange, in as much as a return is expected. Social relationships are built upon reciprocity, with modes of exchange influencing communicative aspects expressing belongingness, but also social obligations. Sahlins (1972) describes social obligations by distinguishing three forms of reciprocity: generalized reciprocity, balanced reciprocity and negative reciprocity. Generalized reciprocity lays the foundations for sharing or giving, with no expectation of a return. This form of reciprocity can be understood as truly altruistic: it is not obligatory to reciprocate, hence the person who gives does not stop doing so if the receiver is not reciprocating. Balanced or symmetrical reciprocity involves direct exchange of equivalent goods or services, hence is demarcated by direct material exchange and promoted by self-interest rather than social obligations. The third form of reciprocity, according to Sahlins (1972), is negative reciprocity, which he describes as getting something for nothing. This form of reciprocity seeks to get the most out of the transaction, with little or no input. In sum, this classification of reciprocity (Sahlins 1972) helps us to reflect on mechanisms of exchange in contemporary society and the absence or existence of reciprocity in sharing as key component of contemporary collaborative consumption (cf. Belk 2014). Considering the wider social organisation of exchange and economic transactions, Widlok (2013) understands reciprocity, also with regards to gift-giving, as being in opposition to commodity exchange. Although trust can be considered primitive regarding economic activity, individuals can reciprocate in different ways and at different levels.

Reciprocity can be positive (reward kindness) or negative (punishment of unkindness) without the participants in collaborative consumption necessarily trusting other participants. Being involved in collaborative consumption does not involve individuals who are always interested in sharing gains with others, whether those others are known to them or not. The question that arises is, what expectations of reciprocity can be found in collaborative consumption: what can I expect to get back? Is it the money in capitalist societies or is it greater self-interest? Solidarity? Love? Expectations differ, depending on how we regard new forms of consumption and consumers, but also new business models in collaborative consumption. It is certainly true that in both cases the environment (sustainable consumption) is taken into account. In new forms of collaborative consumption, changes in social practices may have a greater impact, e.g. not using the car but walking or going by public transport. These changes are associated with new lifestyles that are associated with the ideas of degrowth and living a better life by consuming fewer resources. Hence, the value of sharing and gains at community and environmental level outweigh the economic benefits of using and accessing goods and services (Bardhi and Eckhardt 2012). The meaning and importance of property shifts as the forms of reciprocity diversify.

The positive expectations of contemporary forms of collaborative consumption focus on community, the common good, and solidarity, not necessarily on individual gains. Locating collaborative consumption in the sharing economy, it is useful to refer to Polanyi (1957), who discussed different economic systems from a historical perspective: redistribution, reciprocal and householding systems. From historicizing economic activity, we learn that householding and reciprocity have been and continue to be linked. The give and take of social kinship in the everyday and household context, also understood as sharing (see Belk 2014), can be emphasized in contemporary forms of collaborative consumption. Contemplating the traditional example of gift giving (e.g. Mauss 1923), reciprocity can be understood as a way of making gains not only for the other party but also for the community. The potlatch is an example of reciprocity being a main cause of human behaviour. Historically, we can contemplate the development of (market) exchange and the introduction of money as overcoming communitarian reciprocity. Monetarisation of the economy has more recently increased with the advent of neo-liberal discourses introducing the fetish of monetizing every aspect of social life. Money in neo-liberal societies has become increasingly important, with the need for direct reciprocity disappearing. Even though money is the main medium of exchange in contemporary societies, increasing processes of digitalization allow for reciprocating more broadly, and eventually even accommodate the abolition of capitalist monetary structures. Turning back to the different forms of exchange differentiated by Polanyi (1957), reciprocal forms of exchange in local communities have been based on moral obligations, and these still exist in families and neighbourhood contexts. Contemporary forms of collaborative consumption might change the reciprocal character as the technological platform of collaborative consumption alienates consumers from producers in the same way money does. Accordingly, it is questionable whether the positive expectation in collaborative consumption will be realised and liberate consumers from dependencies or whether collaborative consumption produces new challenges to be detected.

Arcidiacono (2018) empirically analyses in his contribution to this book reciprocity in off-line and online collaborative exchanges of time banking platforms. Time banking can be perceived as a communitarian practice, addressing the relational aspects of reciprocity in digital and physical interaction. In collaborative economies, reputation is the manifested pervasiveness of reciprocity. The boundaries between online activity and offline activity become blurred, while reciprocity is emerging in the ongoing practices shaping everyday life. Time banking is understood as being disruptive and undermining the liberal logic of exchange. It promotes the idea of building relationships and cultivating the sense of belonging. Arcidiacono (2018) distinguishes three kinds of reciprocity: collaboration based on instrumental motivation, true reciprocity that is closer to the potlatch (Mauss 1923), and common pool arrangements that are seen as the 'true and proper sharing practices'. These three forms of reciprocity show the extent to which we may consider collaborative consumption as a key feature for organizing our daily lives. Households often take centre stage in organizing everyday life, but also extend further, with their members playing a role in the economy as workers and generating an income. Indeed, valuing in money and the role of quantifying what is of value often omits everyday practices in household contexts such as the unpaid work of female labour (Fridrich et al. 2014).

With regard to reciprocity and householding, current discussions on collaborative consumption also invite us to revisit the concept of the "prosumer" (Toffler 1980), which reflects the amalgamation of production and consumption in transformative processes such as 3D printing, the open source movement, and investing in solar panels, i.e. producing and consuming simultaneously. Even though such activities may generate monetary income for consumers and emphasize the productive character of consumption practices, the question is whether the theoretical distinction between producer and consumer holds true and whether true reciprocity can be detected. The traditional boundaries of commercial market-oriented conceptualisations of the consumer are blurring (cf. Fridrich et al. 2014). It can even be said that the technology-driven examples of collaborative consumption are only the tip of the iceberg, as mundane and banal activities in everyday household production have become widespread, with reciprocity being a key feature of daily lives of consumers. Yet new business models based on prosumption may, in an unregulated, platform-based collaborative economy, generate social injustice for workers and become obstacles to collective representation instead of being liberating.

The boundaries of work and consumption in the sharing economy and collaborative forms of consuming are analysed in the contribution of Laamanen, Borraz and Islam (2018). They question the liberating effects of collaborative consumption and scrutinize the exploitation of consumers involved in value creation. In their contribution, they link collaborative consumption and production to work issues, and thereby question the role of work in collaborative consumption and the boundaries of the dichotomist thinking. They attempt to answer questions such as: What is work? How are work and the role of the producer reflected in collaborative consumption practices? How are the entrenched roles of capitalist market systems revealed? Work in collaborative consumption spreads beyond contractual labour and is becoming part of the everyday life of consumers in households and families, with reciprocity playing a key role. This helps us to reconsider the power of corporate platforms as more or less invisible and workers' representation as non-existent. Yet, workers' representation should create reciprocity as well as solidarity in collaborative consumption practices.

Indeed, the consumer as producer in household contexts has often been neglected. From a traditional economic viewpoint, consumers buy perfect commodities that are ready to use, for intermediate consumption. In fact, consumers only make concessions, and the commodities are transformed and produced in household processes for fulfilling the needs of consumers in their everyday life (see Fridrich et al. 2014). Examples range from furniture that is bought and has to be assembled at home or ready to eat convenience food waiting in the freezer to be warmed up in the microwave oven. Consumers, therefore, have always been involved in the creation of value. Collaborative consumption helps us to acknowledge the creativity involved in influencing the production process, and the individualization of commodities reveals the interplay between producers and consumers, between planning in the household context, industrial production and intermediate consumption.

Finally, Ingun Krimpstad Klepp and Kirsi Laitala elaborate on the role of reciprocity in different forms of sharing with regard to fashion consumption. Different forms of resource allocation and circulation can be distinguished, such as borrowing, stealing and inheriting. Reflecting on the environmental impact of pieces of clothes worn in family households also hints at the role of reciprocity and distinguishes between intra- and extra-household sharing. Changes involve new business models but also alternative lifestyles that operate based on new technologies, enabling alternative means of sharing. In their understanding consumption includes acquisition, use and disposal of goods and services.

To conclude, this book brings together various perspectives on collaborative consumption, shedding light on empirical and theoretical aspects pertaining to trust and reciprocity. The nuances of social and economic activity in collaborative consumption assist us in considering how social change is moving in the direction of more sustainable consumption. The authors of this book are interested in ascertaining the extent to which collaborative consumption challenges prevailing productionist economic models and their associated growth paradigm (Fridrich et al. 2014), in order to create a more sustainable and liveable future.

References

Arcidiaono, D. (2018). Reciprocity in the Sharing Economy: The Case for Time Banking Platforms. In I. Cruz, R. Ganga & S. Wahlen (eds.), *Contemporary Collaborative Consumption*. Wiesbaden: Springer VS.

Bardhi, F., & Eckhardt, G. M. (2012). Access-based consumption: The case of car sharing. *Journal of Consumer Research*, 39 (4): 881–898. doi:10.1086/666376.

Beck, U. (1998). *World Risk Society*. Cambridge: Polity Press.

Belk, R. (2014). Sharing versus pseudo-sharing in Web 2.0. *The Anthropologist* 18 (1), 7–23.

Berg, J., Dickhaut, J., & McCabe, K. (1995). Trust, Reciprocity and Social History. *Games and Economic Behavior*, 10, Academic Press Inc., 122–142. doi:10.1006/game.1995.1027

Botsman, R., & Rodgers, R. (2010). *What's Mine is Yours: How Collaborative Consumption is Changing the Way We Live*. London, UK: HarperCollins.

Carfagna, L. B., Dubois, E., Fitzmaurice, J. C., Ouimette, M. Y., Schor, J. B., Willis, M., & Laidley, T. (2014). An emerging eco-habitus: The reconfiguration of high cultural capital practices among ethical consumers. *Journal of Consumer Culture,* 14: 158–178, doi:10.1177/1469540514526227.

Fridrich, C., Hübner, R., Kollmann, K., Piorkowsky, M.-B., & Tröger, N. (2017). *Abschied vom eindimensionalen Verbraucher*. Wiesbaden: Springer VS.

Fridrich, C., Hübner, R., Hufnagel, R., Jaquemoth, M., Kollmann, K., Piorkowsky, M.-B., Schneider, N. F., Tröger, N., & Wahlen, S. (2014). Bamberger Manifest für ein neues Verbraucherverständnis. *Journal für Verbraucherschutz und Lebensmittelsicherheit,* 9: 321–326. doi:10.1007/s00003-014-0880-1.

Giddens, A. (1994). Risk, trust, reflexivity. In U. Beck, A. Giddens & S. Lash (eds.), *Reflexive Modernization* (pp. 184–97). Cambridge (UK): Polity Press.

Hübner, R. (2017). Nachhaltigkeits- und Verbraucherforschung. In C. Fridrich, R. Hübner, K. Kollmann, M.-B. Piorkowsky & N. Tröger (eds.), *Abschied vom eindimensionalen Verbraucher* (S. 161–200). Wiesbaden: Springer VS.

Mauss, M. (1923). "Essai sur le don". *Année Sociologique,* 2, 1. Paris: PUF.

Laamanen, M., Wahlen, S., & Campana, M. (2015). Mobilising collaborative consumption lifestyles: a comparative frame analysis of time banking. *International Journal of Consumer Studies,* 39: 459–467. doi:10.1111/ijcs.12190.

Laamanen, M., Barros, M., & Islam, G. (2018). Collective representation on collaborative economy platforms. In I. Cruz, R. Ganga & S. Wahlen (eds.), *Contemporary Collaborative Consumption*. Wiesbaden: Springer VS.

Laitala, K. et al. (2018). Shared use and owning of clothes: Borrow, steal, or inherit. In I. Cruz, R. Ganga & S. Wahlen (eds.), *Contemporary Collaborative Consumption*. Wiesbaden: Springer VS.

Paltrinieri, R., & Spillare, S. (2018). Collaborative Consumption and Trust-Building Processes in the Emerging New Food Economy. In I. Cruz, R. Ganga & S. Wahlen (eds.), *Contemporary Collaborative Consumption*. Wiesbaden: Springer VS.

Piorkowsky, M.-B. (2017). Konsum im Fokus der Alltags- und Lebensökonomie. In C. Fridrich, R. Hübner, K. Kollmann, M.-B. Piorkowsky & N. Tröger (eds.), *Abschied vom eindimensionalen Verbraucher* (S. 73–112). Wiesbaden: Springer VS.

Polanyi, K. (1957). *The Great Transformation*. Boston: Beacon Press.

Ritzer, G. (2014). Prosumption: Evolution, revolution, or eternal return of the same? *Journal of Consumer Culture,* 14(1), 3–24. doi: 10.1177/1469540513509641.

Sahlins, M. (1972). *Stone age economics*. London: Routledge.

Schor, J. B., & Fitzmaurice, C. J. (2015). Collaborating and connecting: the emergence of the sharing economy. In L. A. Reisch & J. Thøgersen (eds.), *Handbook of research on sustainable consumption* (pp. 410–425). Edward Elgar Publishing.

Setiffi, F., & Lazzer, G. P. (2018). Riding Free-Riders? A Study of the Phenomenon of Blablacar in Itlay. In I. Cruz, R. Ganga & S. Wahlen (eds.), *Contemporary Collaborative Consumption*. Wiesbaden: Springer VS.

Slee, T. (2015). *What's Yours is Mine. Against the Sharing Economy*. New York, NY: OR Books.

Toffler, A. (1980). *The Third Wave*. New York: William Morrow.

Vaquero, A., & Calle, P. (2013). The Collaborative Consumption: A form of consumption adapted to modern times. *Revista de Estudios Económicos y Empresariales,* 25, pp. 15–30.

Wahlen, S., & Laamanen, M. (2017). Collaborative Consumption and Sharing Economies. In M. Keller, B. Halkier, T.-A. Wilska & M. Truninger (eds.), *Routledge Handbook on Consumption* (pp. 116–130). Routledge (Routledge International Handbooks).

Wahlen, S. (2018). "Foodsharing": Reflecting on individualized collective action in a collaborative consumption community organisation. In I. Cruz, R. Ganga & S. Wahlen (eds.), *Contemporary Collaborative Consumption.* Wiesbaden: Springer VS.

Widlok, T. (2013) Sharing. Allowing others to take what is valued. *HAU: Journal of Ethnographic Theory,* 3/2: 11–31.

Reciprocity in the sharing economy: the case for time banking platforms

Davide Arcidiacono

Abstract

The chapter is focused on the issue of Time banking and its transition towards digital time deals within the sharing economy paradigm. Time banks usually include an important social and emotional component and are oriented towards the (re)construction of a community in a specific territory, but in the new era of sharing 2.0, the time banks go online and are reinventing this practice. The chapter undertakes a comparative analysis of the results of several pieces of research on time banking online and offline, trying to assess differences between and quality of the new digital time banks. The conclusion discusses the opportunities as well as the limitations and risks in this digital transition.

Zusammenfassung

Dieses Kapitel erörtert Zeitbanken und digitale Zeit dem Paradigma der Sharing Economy folgend. Zeitbanken umfassen in der Regel eine wichtige soziale und emotionale Komponente und sind auf die (Re-)Konstruktion einer Gemeinschaft in einer spezifischen Lokalität ausgerichtet. In der Sharing Economy und der Ära des Teilens 2.0 erfolgt der Austausch in Zeitbanken digital und sie erfinden diese Praxis neu. Das Kapitel untersucht im Rahmen einer vergleichenden Analyse Online- und Offline-Zeitbanken. Es werden Unterschiede und Gemeinsamkeiten durchleuchtet und die Qualität der neuen digitalen Zeitkonten bewertet. Dies hilft, um Möglichkeiten, Grenzen und Risiken dieser digitalen Transformation auszuloten.

Keywords

time banking, reciprocity, sharing economy

I. Cruz et al. (Hrsg.), *Contemporary Collaborative Consumption*,
Kritische Verbraucherforschung, https://doi.org/10.1007/978-3-658-21346-6_2

Reciprocity 2.0: between tradition and innovation

It was over 60 years ago that Polanyi (1944) noted how reciprocity was one of the main forms of economic exchange based on moral obligations and defined by membership of a community. Reciprocity is a form of economic integration based on gift trade that according to anthropologists like Malinowski (1924) and Mauss (1922) was historically the basis of social life. We could say that economic transactions based on reciprocity and gift exchange have historically characterised societies, but such schemes have been marginalised by the strength and pervasiveness of the market form of exchange. As argued by Mauss, originally the main forms of reciprocity were linked with ceremony rituals, and the gift trade was capable of merging communitarian practice and some germinal forms of market trade in the transition from the *gemeinschaft* (community) to the *gesellschaft* (society) (Tönnies 1887).

This concept was used by Polanyi to demonstrate how economic practices were socially embedded and included economic activity with a mix of utilitarian and non-utilitarian aims. He argues how individual profit maximisation through the market exchange mechanism is one of the late forms of economic integration that is historically preceded by other methods, like reciprocity, which excluded the idea of the highest individual gain and the commodification of all the factors of production. According to the reciprocal scheme of exchange, goods and services are produced and traded on the basis of an expectation of receiving other goods and services in a manner and at a time that is reasonable within shared social norms. In this case, the economic action is designed to reaffirm the sense of community from which you expect to receive what is necessary for your existence.

Reciprocity continues to adapt, through the family, kinship, voluntary activity or within other significant relations including those in developed societies, but where it is no longer the predominant form of economic transaction. An immeasurable amount of goods and services continue to be exchanged in reciprocal schemes, and contribute not only to the welfare of their members but also to the functioning of the state and the market. The material and cultural reproduction of the labour force, the amount of assets required by the market, the demand for a variety of services (administrative, social, health, safety) are dependent on how the community institutions (family, kinship, volunteering) organise their activities and integrate reciprocity, market and redistribution in a complementary strategy.

Polanyi predicted that after the expansion of forms of redistribution through the welfare state, it would be necessary for an increase in new forms of reciprocity to compensate and sustain the public redistribution effort (Provasi 2014). Polanyi's prophecy has been realised in the crisis of the welfare state, accelerated by the austerity programme, the burden of the public debt, and combined with market failures in the economic crisis of 2008, which created the right conditions for the rise of a new period of reciprocity practices. What are the main characteristics of reciprocity and why are they relevant to the collaborative consumption phenomenon? How are the aims and features of reciprocal schemes changing in a digital and complex social environment? The main characteristics of the Polanyian concept of reciprocity have often been recalled and quoted by scholars (Ouchi 1980; Bradach et al. 1989; Thompson 2003) but they used the term with a different meaning. Ouchi insisted more on the idea that reciprocity is above all a domestic form of economy, while Bradach had a more extensive conception of reciprocity within inclusive networks and based on trust, whereas Thompson tried to distinguish reciprocity between networks and clans.

Reciprocity is a strong device for enforcing social norms and social cooperation, despite the risk of non-cooperation, as a person who has received a gift is not obliged to give one in return, so gifting is a gamble as reciprocation is not ensured. Simmel (1907) argues that this makes gift trading intrinsically powerful but also conflicting. The reciprocity, based on moral obligation, creates a sense of duty that is difficult to ignore and could create a sense of dependence on the donor, with a high risk of personal tension, as refusing a gift is offensive and is a way of harming the relational bonds that the exchange aims to create. At the same time, reciprocating a gift immediately is a way of communicating a personal sense of discomfort with the relational and moral obligation. As Mauss argued, reciprocity is a dynamic balance between gratuitousness and individual interest such as between symmetry and asymmetry. Reciprocity does not require certainty about time or value equivalence, which leads to an avoidance of the endeavour to cooperate – a free riding attitude (Olson 1965).

There is some empirical evidence (Fehr and Gächter 1998) that demonstrates how the role of free riding in collective action might be overestimated, because the idea of material calculation could be nuanced in a reciprocal scheme where if "others are behaving cooperatively, individuals are moved by honour, altruism, and like dispositions to contribute to public goods even without the inducement of material incentives. When, in contrast, they perceive that others are shirking or otherwise taking advantage of them, individuals are moved by resentment and pride to retaliate. In that circumstance, they will withhold beneficial forms of cooperation even if doing so exposes them to significant material disadvantage [...] By the same token, individuals who lack faith in their fellows can be expected to resist contributing to public goods, inducing still others to withhold their cooperation as a means of retaliating. In this self-sustaining atmosphere of distrust, even strong (and costly) regulatory incentives are likely to be ineffective in promoting desirable behaviour" (Kahan 2002, p. 2). Kahan offers some interesting insights on the peculiarity of the reciprocity logic: it sees economic and social actors as moral and emotional reciprocators more than simply egoistic wealth maximisers or even free riders. If in conventional theory of collective action, defection or free-riding are the dominant strategies, in reciprocity, individuals prefer to contribute if they believe others are inclined to contribute, or they do the opposite if they believe others are inclined to be free riders, in a sort of multiple equilibrium strategy of collective action.

The cooperative endeavour is based more on trust than on an efficient incentive system, and only a small fraction of the population consists of committed free-riders, just as a small fraction also are dedicated co-operators who love to contribute in any case. The power of reciprocity is thus based on the cohesion of the group, because it does not require some to contribute more than others in the group, so "the individual has obligations not to society, but to any group of individuals from whose effort he derives benefits" (Sudgen 1984, p. 775). However, the risk of "reciprocity traps" cannot be discounted, because reciprocal transaction schemes may not be automatically inclusive, equal and fair towards the rest of the society. We need to make the distinction argued by Bruni (2006) between cautious reciprocity, where my cooperative attitude is subordinated to the cooperative attitude shown by others, brave reciprocity, where my decision to cooperate is not subordinated to the attitude of others but where I could reverse my decision if nobody else cooperated, and unconditional reciprocity, where I cooperate anyway and despite the behaviour of others. The majority of people cooperate conditionally on the availability of others to contribute. According to recent studies (Buchan et al. 2000), the levels of trust and reciprocation may vary across countries, so cultural variables and the socio-institutional environment are relevant to reciprocity schemes in economic transactions.

As argued before, reciprocity schemes are changing in the complex digital and social environment. Today's Web 2.0 has increased new forms of reciprocity that are defined using different expressions: collaborative consumption (Botsman and Rogers 2010), circular economy (Morace 2014), and commons-based peer production (Benkler and Nissembaum 2006). Collaborative consumption practices propose the gradual replacement of private ownership with distributed access rights (de-ownership and disintermediation of the exchange). They can be seen as a "natural" development or even as an overcoming of capitalism and its traditional conception of consumption as a destructive process (Rifkin 2014).

Collaborative consumption is a transformative and disruptive paradigm because it is based more on the relational value of consumption and its balanced mix of tradition and innovation where reciprocal schemes of transactions and relations represent a connective tissue between digital and physical interactions. In a society that is increasingly digital, the possibility of sharing and coordinating reciprocal schemes is increased, as we rediscover the value and relevance of a model of transactions that we imagined was more a symbol of pre-modern society. At the same time, this rise in reciprocity raises new considerations. First of all, it operates within new forms of communities, particularly online communities, that interact through the mediation of digital social platforms. This increases the capacity for new interactions and opportunities. These communities do not share a common identity, are not rooted in a specific territory and are not inspired by the same values or even socialized to the same social rules, so we might speak of "imagined communities" (Anderson 2006).

Another interesting feature is the increasing role of digital reputation systems in the generative process of trust based on these new reciprocal schemes. Some scholars have argued that digital reputation ratings could push a more 'ethical' and consumer/society driven economy (Arvidsson and Pietersen 2009), others mention the risk of promoting a stronger selfhood (Papacharissi 2011). As Hearn (2016) puts it, "reputation is a cultural product, and, as such is conditioned by its mode of production. This mode of production is generally marked by the perennially exploitative relations between labour and capital [...] The claim that the interactive feedback represented in measured forms of 'reputation' will inevitably produce socially ethical behaviour seems far-fetched in a world where market logics and values remain dominant and cynical forms of self-production prevail. Individuals generally craft reputation via the self-brand because they hope this work will eventually find its realisation in the general equivalent – money; [...] As reputation seekers in the reputational economy, we may be nodes in the new distributed means of production as producer, product, and consumer, but we do not even begin to control the means of our own distribution" (pp. 421–435). In this sense, digital reputation systems amplify the pervasiveness of reciprocity but at the same time seem to subordinate it to a market logic.

The growing importance of reciprocal forms of exchange within the collaborative consumption paradigm could also be described as a mobilising strategy of the so-called lifestyle movements (Haenfler et al. 2012), as a form of resistance (De Certau 2001) of the citizen-consumer (Arcidiacono 2013) in order to re-embed the economic practice through a proactive use of digital technologies promoting new connective social action (Bennett and Segerberg 2012). These ideas include the new forms of reciprocity as a way of promoting social innovation (Hamalainen et al. 2007) or of responding to the so-called tragedy of the commons (Hardin 1968; Ostrom 2002). However, it is evident that it is not possible to conclude that a true collaborative movement actually exists at the moment (Schor 2014).

Aware of the new complexity introduced in reciprocity 2.0, Pais and Provasi (2015) have tried to rescale reciprocity practices in the sharing economy by distinguishing between collaboration, true reciprocity and common-pool arrangements (p. 360). If collaboration is guided exclusively by instrumental motivations, somehow more akin to market interactions, true reciprocity is closer to the traditional gift economy, which does not exclude self-interest but coexists with the acceptance of the risk of non-cooperation; finally, common pool arrangements are the true and proper sharing practices, based on the development of a common sense of belonging and a process of mutual recognition that spans the identities of those who exchange goods and services.

The rest of the chapter will analyse how reciprocity changes within a specific practice such as time banking, it will also look at the different configurations it assumes when moving from off-line to online collaborative exchanges. Time banks (TB) are one of the most illustrative cases of the value of reciprocity in modern societies, historically forerunners of collaborative practices well before the current debate on the sharing economy and collaborative consumption.

A question of time: sharing and reciprocity of time resources in contemporary society

Time is one of the major concerns of our life (Adam 1990) because decisions on the allocation of this resource are always the result of a compromise between natural and social norms. In particular, in Western societies there has been a transformation of social timing along fundamental lines; speeding, measurement, expansion and commodification (Gershuny 1993). In the post-industrial phase, the meaning of time varies, in particular due to the disappearance of the traditional separation between working and non-working hours. There is a new "social rhythm", that risks a qualitative and quantitative depletion of temporal resources (Bauman 2002), generated by the recurring dynamic of a negative overlap between work and private life, and the overlapping of different layers of time.

Despite the consequences of these transformations, time continues to be a necessary resource in the process of social reproduction, even with increased pressure for its configuration as a "market object" (Thompson 1969; Gasparini 2009), thanks to the demand for services aimed at addressing the acceleration processes, liquefaction and fragmentation of social timing. Social times were affected by neo-liberal de-regulation processes, in particular, the deregulation of the labour market, and new forms of social inequality have increased irregularities in the individual or collective ability to "liberate" and "qualify" their own temporal resources.

When faced with such transformations people try to generate practices and strategies to reconnect to the fundamentals of social times such as time banks, which is a tool for mutualism and reciprocity within specific and territorially defined communities (residents of a neighbourhood or the human resources of a company). The system is configured as a proximity service that pools the scarce time resources of each person and redistributes them to support the needs of a member of the community. Through the time bank, the members of a community exchange services, the value of which is defined by the amount of time required, so that time acts as a type of currency. After the first experiments, time banking was discussed as a move away from individualistic liberal discourse that could overcome an unequal and exploitative system of exchange. Time banking services were the first good example of a cooperative production between the private and public spheres, anticipating much in the current debate on the construction of new participative and collective endeavour for the commons.

There are different models of organisation of a TB (Dittmer 2013, p. 6):

a) traditional institutions in education, health, social care or justice sectors develop specific programmes rewarded in time credits;
b) person-to-agency time banks, run by a public or third sector actor, as a tool for achieving social goals by changing the incentive structure of a targeted group, and
c) agency-to-agency time banks to assist the sharing of capacity between organisations.

The idea was first developed by Professor Edgar Cahn of the London School of Economics (LSE) in 1986: "Help a neighbour and then, when you need it, a neighbour – most likely a different one – will help you. The system is based on equality: one hour of help means one-time dollar, whether the task is grocery shopping or making out a tax return […] Credits are kept in individual accounts in a 'bank' on a personal computer. Credits and debits are tallied regularly. Some banks provide monthly balance statements, recording the flow of good deeds." (New Economic Foundation 2001, pp. 5–6).

Time sharing activity is based on a variety of services that have been exchanged within the hour-per-an-hour scheme, from childcare to computer lessons, plumbing or gardening. The experience of time banking in the United States of America (USA) became even more specialised and in some cases took on the form of a self-help system, or a protocol of social rehabilitation. In some American prisons, prisoners use this system for earning time credits which are spent to stay in touch with their children, or people with depression exchange time credits for looking after older local people. Analysis of the American case shows the majority of part-time credits accumulated are mostly never spent, providing evidence of how the system promotes altruistic activities and voluntary work, so those who offer time often give more than they want to receive. The American experience spread rapidly overseas, and Japanese and Chinese public administrations have adopted time banks systems as a way of providing support to their rapidly ageing populations, so their system is integrated into social service and volunteering policies.

In Europe, the origins of this practice date back to the 1990s, during a time when there was an increase in harsh socio-political debate about social care. Italy, the United Kingdom (UK) and Spain (Amorevole et al. 1998; Williams et al. 2003) were among the European countries that stressed the need for more experiments with local time banking systems. The Local Exchange Trading Scheme (LETS) in the UK was one of the most important experiments using this form of exchange and led to the development of the practice of time banking there. In less than five years, from 1998 to 2003 there were more than 130 time banks in the UK, starting in the county of Gloucestershire, which was in the vanguard of innovative practice. In the UK, time banks were quickly implemented as a way of redesigning social care services (for example, Time for Childcare in the Cotswolds and Leicester, Rushey Green Group Practice in a surgery service) re-embedding the human side of an "anonymisation" started by the push to deregulation and privatization that began in the 1980s and turned citizens into customers. However, TBs are also a "social capital builder" (Boyle 2003, p. 254) which help to create social capital and collective responsibility as a tool for inclusive and relational markets within resilient local communities (Schroeder et al. 2011).

Time exchanges usually include an important social and emotional component and are oriented towards the (re)construction of a sense of community belonging in a specific territory. According to a survey of the Lets experience, 55 % of respondents confirmed they developed their network of friends and 75 % developed self-help networks (Williams et al. 2001). The average size of traditional time banks is approximately 60 members, but there are many even bigger ones such as *STAP* in Helsinki (with over 3,000 members including collectives and organisations), or *Timebank.cc* in Netherlands (with more than 1,200 members).

In many cases the time banks are composed of local residents, mostly women, unemployed and from lower-income households, and also including people not traditionally committed to volunteering activities (Seyfang 2003; 2004). Other authors (North 2003; Dittmer 2013) confirmed the increasing role of TBs as a collaborative tool that usually operates in deprived areas where there has been a failure in unskilled personal services and as a way of developing local self-reliance. Even considering the motivational profile of the members of the TBs, recent studies (Laamanen et al. 2015) demonstrated the centrality of cultural and economic transformation within society, while highlighting different motivations for involvement. Some members want to move towards a non-monetary economy, others propose a more political and economic realignment of local decision-making, while others simply seek to develop forms of friendship and care within their neighbourhood.

Seyfang (2004) distinguishes between the internal and external obstacles to the functioning of TB's. The internal obstacles are linked to members' behaviour, and many participants consider time banking exchange to be like a voluntary activity and give more of their time but are reluctant to ask for help for themselves, so offers of time are more numerous than demands for time, limiting the range of choices and potential interactions. Some members who have less direct contact with the TB's become increasingly less involved because they feel cut off, and need to be empowered via the mechanism of informal social gatherings and strong communication. These obstacles could explain the limited level of time transactions in some of the experiences analysed, which also limits the building of a shared and inclusive sense of community (Dittmer 2013; Lee et. al. 2004).

The external obstacles are linked to the dependence of TB's on grant funding to support the spaces and human resources (e.g. the time brokers) required for their operation. Guaranteeing the long-term sustainability of the TB promotes the most benefits in terms of improved relations, increasing solidarity and community building. The analysis in UK and Sweden shows that tax exemptions or other benefits promoted by regulation are relevant to the success and expansion of this system of exchange (North 2003; Molnar 2011).

Objectives and method

This chapter examines how motivations and exchange practices change in the transition from the offline to the online system, with its associated impact on reciprocal transaction mechanisms. The data collected examined the transactions of one of the most important digital time banking platforms currently in operation, Time-Republik (TRK) (Arcidiacono 2015). TRK was created in 2013 with the aim of bringing the bank's system of time into a global, social and digital environment. The research is based on a net-nographic approach (Kozinets 2009), and the use of the mystery shopping technique (Turner 2007). In a logic of triangulation (Denzin 2006) we have combined the qualitative data collected through the direct experience of interactions and transactions within the platform with the quantitative data on platform use collected through a survey of users (N = 238) in 5 countries. The self-filling questionnaire was sent out with CAWI methodology – to all platform users – and the result is thus, in part subject to a self-selection bias, revealing which respondents may represent the most active component of TRK.

Time banking 2.0: how does reciprocity change?

New digital technologies increase and strengthen the power of reciprocity within the umbrella concept of a sharing economy. Since 2008 a new kind of TB has emerged, the Digital Time Bank (DTB) such as *Cronobank, Bliive,* and *Time-Republik.* It is essential to analyse how this transformation influences reciprocal transactions, to evaluate this new phase of development in time banking and to test the real "re-socialising" capacity of these new forms of collaborative consumption practices. DTBs would eliminate or decrease the impact of external obstacles such as economic sustainability by creating a cheaper model, one less dependent on public funding and benefits, as happened to some of the traditional TBs. Furthermore, DTBs could potentially increase the intensity of interactions through a totally disintermediated settlement model, where people negotiate time credits and transactions directly without the need for a time broker.

Traditionally the territory of time banks was confined to a neighbourhood or a more limited territorial/organisational space (closed membership), but in the new era of sharing 2.0 time banks may increase the space for possible interactions/transactions. DTBs are potentially less bounded by spatial limits, which opens up the opportunity for transactions beyond territorial borders, and increasing the number of time credits exchanged through the viral power of the web (open membership). Time-republik, for example, has more than 30,000 users who exchange more than 100,000 competencies in 110 countries, while Beliive offers over 124,470 time credits.

The activities shared on TRK are mostly related to arts, media and design. Among the most commonly shared services are computer-related assistance, translations, photography, graphic design, web development and social media management (Arcidiacono 2015). They are mostly services and activities that, thanks to the support of the Internet, can be exchanged quickly and virtually with little possibility of entering into direct contact with other people. Deals are concluded online and do not require any physical proximity, with the majority of interactions taking place among people in the same municipality, region or state. The geographical origin of those seeking help continues to have a decisive influence, as in the traditional TBs, so it could be concluded that DTBs act as a glocal platform capable of activating potential connections between users located in every corner of the planet, but with concrete operability concentrated at the local level. So the local dimension of time exchange does not disappear at all, its borders are simply enlarged. On the one hand the Internet makes the extension of networks of contacts on a global scale possible, on the other, national languages and cultural codes continue to play a part in contributing to a sort of "regionalisation" of global cyberspace which contrasts with the mythology of the network as a space without borders.

DTBs users are increasingly "digital", highly educated and young, the so-called millennials, with a high rate of gender homophily (Pais and Dal Maral 2016). This is also true for the platform analysed. The user-profile of DTBs is different and more heterogeneous than that of TBs, linked by support for social activism and strongly motivated by ideological drivers. The DTBs show a different pattern of mobilisation, with a mix of public and private interests. In this the users of a DTB are classified into four ideal-types based on two fundamental dimensions (the intensity of time transactions and the level of relational interactions online):

- *the idealists* represent a niche in the platform, and are the most available and oriented to an economy of sharing who have consciously chosen to donate their time. Their motivations are primarily ethical, environmental sustainability and the desire to contribute to the spread of altruistic practices and values. They believe that money is not the only way to attribute economic value to a product or a service;
- *the pragmatists* use the system in a practical and more opportunistic way. These users give time because they see the platform as a means of achieving their own personal purposes and have relational exchanges that are strictly limited to their needs;
- *the professionals* value the time platforms as a tool for self-promotion and professional visibility. They participate selectively, but are interested in building relationships, as they want to build a large network. They are mostly communication or digital and information technology professionals;
- *the spectators* are the most numerically significant and represent an unexpressed "potential". They join a DTB platform out of curiosity or for exploratory reasons but have a timid, private and more suspicious approach and tend not to deal.

Time banks seem to have a chameleonic nature, with different types of organisation and management models that are not always egalitarian or practicing what they promote. In fact, in cases like Time-republik, they are a for-profit organisation with a very different framework and approach to what we see in traditional and local TBs. Pais and Del Moral (2015), in a recent study on digital time banking platforms, pointed out some paradoxes with for-profit DTBs. The first is how the goal of social cohesion can be promoted, in the light of the entrepreneurial aims of the majority of the well-known DTBs. In the case analysed, the main strategy is a sort of freemium account which distinguishes between individuals and collective actors (private companies, associations, foundations, etc.). The second paradox is the relationship between the global perspective of the DTBs and the importance of the local/territorial dimension. In the case analysed, this paradox is addressed by empowering and strengthening relations between the platform and local traditional time banks, even though this has not been successful, because the Italian Association of Time Banks has not accepted Time-republik as a member. Another relevant paradox is the types and aims of their transactions. It is argued that the growth of time banking in the digital environment coincided with the more recent transformation in work relations due to the use of Information Communication Technology (ICT) (Scholz 2012; Huws 2013; Sandurarajan 2016), with fuzzier boundaries between work and non-work because "a generation already primed to accept the interpenetration of 'fun', 'education' and the normal business of life online was ill-equipped to dispute such slippages in relation to boundaries between these activities and 'work' in other spheres" (Huws 2013, pp. 4–5). In this scenario, time banking transactions are a good example of prosumption and of the blurring of boundaries between consumption, production and digital labour (Dujarier 2009). In this, DTBs are similar to a marketplace for young talented professional-amateurs (ProAm), instead of a system of solidarity and mutual help.

Close analysis of the demands for and offers of help in some DTBs shows a similarity with some job applications often found on major job boards or in professional social networks. Some offers communicate their skills in an "emotional and personal" way, as in a covering letter for a job application, others try to attract attention by using a "commercial" and catchy style. The majority of talent shared is linked to the digital world of computers, because these skills are easier to share digitally and because there are a lot of young professionals and highly skilled computer users, often called the "digital proletariat" or "digital nomads" who are part of a wide and connected community, who take care of their digital reputation and build ties that might be useful in their careers. In addition, many of the skills exchanged are related to jobs that are not fully recognised and codified on the market, so through DTBs people are trying to legitimise their value and professional contribution as, for example, in the case of a *pet walker* or a *life coach*. The identity of DTBs seems to merge into a crowdsourcing platform such as E-lance, Freelancer, and O-desk. This confirms Marcel Mauss's conception of the gift exchange that is not only for free but also based on personal and individualistic interests. Gift and market exchanges are not two opposites in a Polanyian sense, but two sides of the same coin.

There is some research evidence that highlights how time banking services do not eliminate the risk of conflicting transactions and access discrimination and operate in a more uncertain regulatory framework in terms of user protection. In particular, Dubois et al. (2014) found that some members with high cultural capital prefer trading with others in that social stratum (social class homophily). Network analysis seems to confirm how in time platform banking there were very limited reciprocal links, and most of the relations remain quite limited to unidirectional dyadic relationships. These links are often characterised by a one-way relationship between two users, which implies a clear division between the role of requester and the role of performer. Relationships are strongly conveyed and restricted to a few individuals and a limited and instrumental relational approach emerges among the users, where every attempt at contact without an exchange transaction of time is considered spam, abusive, or perceived as "inappropriate", resulting in low generation of relationships inside or outside the platform. This low importance of the relational dimension demonstrates how far we are from the construction of that sense of community which characterised the utopian perception of the sharing economy or the solidarity of offline TBs. Using a term elaborated by Micheletti (2003), we have a *collective individualised action,* based on a form of cautious reciprocity in the Polanyian sense. Anderson's (1996) definition of *imagined communities,* established by the spread of the social but not sufficient to build a concrete sense of belonging, appears acceptable in this connection. The mix of personal gratification and indirect utility explains the propensity to exchange, but with the conscious intention that the relationships remain distant and impersonal, and thus less risky.

Concluding remarks

This chapter explores the results of research into online and offline time banking, and has sought to assess the differences in the quality of collaborative practices in TBs and DTBs. Collaborative consumption practices like time banking present a huge opportunity to re-embed economic exchange and the relational value of "goods", but this disruptive power is linked to the model of governance of these platforms, and the role of the territory is still relevant as a guarantor and a facilitator of this collaboration model. The territorial homogeneity of deals and interactions explains the need to increase the synergies between online and offline as a key strategy.

The research analysed seems to suggest there had been a careful examination of these practices, but it was limited because it related to single cases. We need to demythologise or avoid uncritical judgement of collaborative consumption practices as a progressive or innovative tool. The exchange model seems to be disruptive and undermines the liberal logic of the power of the market, with some motivations and practices seemingly aimed at playing dynamics very similar to the market scheme instead of safeguarding the fundamental value of time or building relationships and creating a sense of belonging. Mauss's conception of how the gift economy could not be separated from its instrumental purposes is confirmed. When the object of sharing is not goods/physical objects but performance and skills, the process of sharing can also be used as a mechanism to meet the supply and demand of labour, based on disintermediation and personal branding strategies.

In some cases, we are far from the romantic idea of a happy and interactive community, but instead have a narcissistic concept of sharing that helps to explain the use of the platform as a channel for personal reputation management. The relational and communitarian dimension is likely to be partially devalued where there is interaction but there is no relationship (Franchi and Schianchi 2011), building possible "traps of reciprocity" that foster the birth of exclusive rather than inclusive networks. The digital reputation system lacks transparency and control and is insufficient to promote quality and relevant bonds or enhance the role of transactions between people who do not know each other or belong to the same territory.

References

Adam, B. (1990). *Time and Social Theory*. Polity Press, Cambridge.

Amorevole, R., Grisendi, A., & Colombo, G. (1998). *La banca del tempo. Come organizzare lo scambio di tempo: i valori, i principi, i soggetti.* Angeli, Milano.

Anderson, B. (2006). *Imagined Communities.* Verso, London.

Arcidiacono, D. (2013). *Consumatori Attivi. Scelte d'Acquisto e Partecipazione per una Nuova Etica Economica.* Milano: Franco Angeli.

Arcidiacono, D. (2015). *Time-banking on line: between sharing economy and personal branding strategies.* In ESA 2015 Conference Book of Abstracts, Prague.

Arvidsson, A., & Peitersen, N. (2009). *The Ethical Economy.* Online version: http://www.ethicaleconomy. com/info/book.

Bauman, Z. (2000). *Liquid Modernity*. Polity Press, Cambridge.

Benkler, Y., & Nissembaum, H. (2006). Commons-based Peer Production and Virtue. *The Journal of Political Philosophy,* 14: 394–419. DOI: 10.1111/j.1467-9760.2006.00235.x.

Bennett, W.L., & Segerberg, A. (2012). The Logic of Connective Action. *Information, Communication & Society,* 15: 5, 739–768.

Botsman, R., & Rogers, R. (2010). *What's Mine Is Yours: The Rise of collaborative Consumption.* Londra: Harper Business.

Boyle, D. (2003). The New Mutualism and the Meaning of Time Banks. In *LOCAL ECONOMY,* 18: 3, 253–270.

Bradach, J., & Eccles, R. (1989). Price, authority and trust: from ideal types to plural forms. *Annual Review of Sociology,* 15: 97-118.

Bruni, L. (2006). Reciprocità. Dinamiche di cooperazione, economia e società civile, Bruno Mondadori, Milano.

Buchan, R.N., Croson, R.T., & Dawes, R.M. (2000). Direct and Indirect trust and reciprocity, working paper in web.mit.edu/sloan/osg-seminar/pre_02_docs/Indirect.doc.

Cella, G. P. (1997). *Le tre forme dello scambio. Reciprocità, politica, mercato a partire da Karl Polanyi*. Il Mulino, Bologna.
De Certau, M. (2001). *L'invenzione del quotidiano*. Edizioni Lavoro: Roma.
Denzin, N. (2006). *Sociological Methods: A Sourcebook*. Aldine Transaction.
Dittmer, K. (20139. Local currencies for purposive degrowth? A quality check of some proposals for changing money-as-usual. *Journal of Cleaner Production*, 54, 3–13.
Dubois, E. A., Schor, J. B., & Carfagna, L. B. (2014). New cultures of connection in a Boston time bank. In J. B. Schor & C. J. Thompson (eds.), *Sustainable Lifestyles and the Quest for Plenitude: Case Studies of the New Economy*. Yale University Press: New Haven.
Dujarier, M. A. (2009). *Il lavoro del consumatore. Come co-produciamo ciò che compriamo*. Egea: Milano.
Fehr, E., & Gächter, S. (2002). Strong reciprocity, human cooperation, and the enforcement of social norms. *Human Nature*, 13: 1, pp. 1–25.
Franchi, M., & Schianchi, A. (2011). *Scegliere nel tempo di Facebook. Come i social network influenzano le nostre preferenze*. Roma: Carocci.
Gasparini, G. (2009). *Tempi e ritmi nella società del 2000*. Milano: Franco Angeli.
Gershuny J. (1993). *L'innovazione sociale. Tempo, produzione e consumi*. Cosenza: Rubettino Editore
Haenfler, R., Aenfler, B. J., & Jones, E. (2012). Lifestyle Movements: Exploring the Intersection of Lifestyle and Social Movements.*Social Movement Studies*, vol. I, 1–20
Hamalainen, T. J., & Heiskala, R. (eds.) (2007). *Social Innovations, Institutional Change and Economic Performance*. Cheltenham: Edward Elgar.
Hardin, G. (1968). The Tragedy of the Commons. *Science*, 162: 1243–1248.
Hearn, A. (2016). Structuring feeling: Web 2.0, online ranking and rating, and the digital 'reputation' economy, in Ephemera. *Theory & politics in organization*, 10, 3/4: 421–438
Huws, U. (2013). Working online, living offline: labour in the Internet Age. *Work Organisation, Labour and Globalisation*, 7, 1: 1–11.
Kahan, D. M. (2002). *The Logic of Reciprocity: Trust, Collective Action, and Law*. John M. Olin Center for Studies in Law, Economics, and Public Policy Working Papers. Paper 281.
Kozinets, R. V. (2009). *Doing Ethnographic Research Online*. Londra: Sage.
Laamanen, M., Wahlen, S., & Campana, M. (2015). Mobilising collaborative consumption lifestyles: A comparative frame analysis of time banking. *International Journal of Consumer Studies*, 39 (5), 459–467.
Lee, R., Leyshon, A., Aldridge, T., Tooke, J., Williams, C. C., & Thrift, N. (2004). Making geographies and histories? Constructing local circuits of value. *Environment and Planning D: Society and Space*, 22: 4, 595–617.
Malinowski, B. (1922). *Argonauts of the Western Pacific. An Account of Native Enterprise and Adventure in the Archipelagoes of Melanesian New Guinea*. New York: Dutton.
Mauss, M. (1924). Essai sur le don. Forme et raison de l'échange dans les sociétés archaïques. *L'Année Sociologique*, I: 30–186.

Micheletti, M. (2003). *Political Virtue and Shopping. Individuals, Consumerism and Collective Action*. London: Palgrave.
Molnar, S. (2011). Time is of the essence: the challenges and achievements of a Swedish time banking initiative. *International Journal of Community Currency Research* 15: A, 13–22.
Morace, F. (2015). *Crescita felice. Percorsi di futuro civile*. Milano: Egea.
North, P. (2003). Time banks: learning the lessons from LETS? *Local Economy*, 18: 3, 267–270.
Olson, M. (1965). *The Logic of Collective Action*. Harvard University Press, Cambridge.
Ostrom, E. (2002). *The Drama of the Commons*. Washington: National Academies Press.
Ouchi, W. (1980). Markets, bureaucracies and clans. *Administrative Science Quarterly*, vol 25: 129-141.
Pais, I., & Del Moral, L. (2015). Collaborative Economy and the Digitalization of Timebanking: Opportunities and challenges. *Studi di Sociologia*, n. 19: 3–21.
Pais I., & Provasi, G. (2015). Sharing economy: a step towards 're-embedding' the economy?. *Stato e Mercato*, 105: 347–377.
Pais I., & Del Maral, L. (2016). Feminization of Labor, defeminization of Time Banks: digital time banking and unpaid virtual work. In *International Journal of Media and Cultural Politics*, forthcoming.
Papacharissi, Z. (2011) (ed.). *A Networked Self: Identity, Community, and Culture on Social Network Sites*. Routledge, New York.
Polanyi, K. (1944). *The Great Transformation*. New York, trad. Italiana La grande trasformazione. Torino: Einaudi, 1974.
Provasi, G. (2014). Dono, reciprocità, legame sociale. La lezione sempre attuale del "Saggio sul dono" di M. Mauss. In *Iride*, XXVII, 237–294.
Sandurarajan, A. (2016). *Crowd Capitalism*. The MIT Press, New York.
Scholz, T. (2012) (ed.). *Digital Labor: The Internet as Playground and Factory*. Routledge, London.
Schor, J.B. (2014). Debating sharing economy. In *Great Transition Initiative*, http://www.greattransition.org/publication/debating-the-sharing-economy.
Schor, J.B., & Fitzmaurice C.J. (2015). Collaborating and Connecting: The emergence of the sharing economy. In L. Reisch & J. Thogersen (eds.), *Handbook on Research on Sustainable Consumption*. Cheltenham, Edward Elgar.
Schroeder, R.F.H., Miyazaki, Y., & Fare, M. (2011). Community currency research: an analysis of the literature. *International Journal of Community Currency Research*, 15 (A), 31–41.
Seyfang, G. (2004). Time banks: rewarding community self-help in the inner city? *Community Development Journal*, 39 (1), 62–71.
Simmel, G. (1907). *Philosophie des Geldes*. Duncker & Humblo, Leipzig.
Sudgen, R. (1984). The supply of public goods through voluntary contribution. *The Economic Journal*, 94, pp. 772–787.
The New Economics Foundation (2011). *Time Banks. A radical manifesto for the UK*. RAP, London.

Thompson, P. (1967). Time, Work-Discipline, and Industrial Capitalism. *Past & Present*, 38, pp. 56–97.
Thompson, G. F. (2003). *Between hierarchies and markets: the logic and limits of network organization*. Oxford University Press, London.
Tönnies, F. (1887). *Gemeinschaft und Gesellschaft*. Leipzig: Verlag di Fues.
Turner, H. (2007). Mystery Shopping. In *AaVv, Market research Handbook*. Chichester: Wiley & Son.
Williams, C. C., Aldridge, T., & Tooke, J. (2003). Alternative exchange spaces. In R. Lee, A. Leyson & C. C. Williams (eds.), *Alternative economic spaces*. Sage, London: 151–167.
Williams, C. C., Aldridge, T., Lee, R., Leyshon, A., Thrift, N., & Tooke, J., (2001). Bridges into work? An evaluation of Local Exchange and Trading Schemes (LETS). *Policy Studies*, 22: 2, 119–132.

Collective representation on collaborative economy platforms

Mikko Laamanen, Marcos Barros and Gazi Islam

Abstract

The promise of collaborative consumption lies in its potential to promote transparent and participatory communal exchange relations as well as increasing the utilization rate of micro-level capital through sharing with those in need of such capital. The potential of collaborative consumption is supported by technological innovation largely based on platforms of exchange and provision, both in grass-roots efforts and in business ventures. Concentrating on the latter, we argue that the success of collaborative economy business platforms coincides with the fluidity between production and consumption. The differentiation between work and consumption is seemingly thinning: consumers offer their input and resources to organizational processes of production, and consumers' tangible and intangible resources – their possessions and skills – become lucrative and accessible on platforms. Consequently, this challenges traditional definitions, conditions, and sites of work and worker representation, as pinning down what counts as work, and who engages in work becomes increasingly difficult. In this chapter, we address the dearth of research on collective representation in collaborative economy business models. We focus on the ability of collaborative economy platforms to harness working consumers' resources and willingness to work in settings that they do not primarily consider work. We approach the question: What are the possibilities and forms of collective representation in the collaborative economy? Using a synthesizing approach, we illustrate the struggle between individual and collective interests in the relational dynamics of the collaborative economy and theorize on the potentiality for any collective representation in this context.

I. Cruz et al. (Hrsg.), *Contemporary Collaborative Consumption*,
Kritische Verbraucherforschung, https://doi.org/10.1007/978-3-658-21346-6_3

Zusammenfassung

Das Versprechen des kollaborativen Konsums liegt in seinem Potential, transparente und partizipative gemeinschaftliche Austauschbeziehungen zu fördern. Dadurch kann die Nutzungsrate von Mikrokapital mit denjenigen erhöht werden die dieses Kapital benötigen. Das Potenzial des kollaborativen Konsums wird durch technologische Austausch- und Bereitstellungsplattformen unterstützt, welche sowohl in Grasswurzelinitiativen als auch in unternehmerischen Geschäftsmodellen vorzufinden sind. Dieses Kapitel konzentriert sich auf Letzteres und argumentiert, dass der Erfolg von gewinnorientierten Plattformen in der kollaborativen Ökonomie die Fluidität zwischen Produktion und Konsum fördert. Die Unterscheidung zwischen Arbeit und Konsum scheint sich zu verflüchtigen: Konsumenten stellen materiellen und immateriellen Input also Ressourcen für organisatorische Produktionsprozesse auf Plattformen bereit, um ihren Besitz und ihre Fähigkeiten lukrativ zugänglich zu machen. Folglich stellt dies die traditionellen Definitionen, Bedingungen und Arbeitsorte sowie die Arbeitnehmervertretung vor neue Herausforderungen. Dementsprechend wird es schwieriger zu umschreiben was als Arbeit zählt und wer sich an Arbeit beteiligt. In diesem Kapitel befassen wir uns mit dem Mangel an Forschung zur kollektiven Repräsentation in Geschäftsmodellen der kollaborativen Ökonomie. Wir fokussieren uns auf die Fähigkeit von Plattformen kollaborativer Ökonomien, die Ressourcen und die Bereitschaft der arbeitenden Konsumenten zu nutzen, um in Situationen zu arbeiten, welche sie nicht primär als Arbeit betrachten. Wir nähern uns der Frage: Was sind die Möglichkeiten und Formen kollektiver Repräsentation in der kollaborativen Ökonomie? In einem synthesierenden Ansatz illustrieren wir den Kampf zwischen individuellen und kollektiven Interessen in der Beziehungsdynamik der kollaborativen Ökonomie und theoretisieren die Potentialität für jede kollektive Repräsentation in diesem Kontext.

Keywords

collective representation, work, collaborative economy

Introduction

The promise of collaborative consumption lies in its potential to promote transparent and participatory communal exchange relations (Botsman and Rogers 2010). Proponents of collaborative consumption and economy stress the increasing utilization rate of micro-level capital through sharing with those in need of such capital. Sharing thus meets human needs while promoting social and environmental sustainability. Sharing privately owned means of production within the community represents a new form of socialism and mutualism in distribution and equality. In this vision, communities build reciprocities between participants, expand skills to improve local well-being, and increase community cohesion and resilience. In the practice of collaborative consumption, ownership does not transfer, but participants access "humanised" and "authentic" services. The potential of collaborative consumption is supported by technological innovation, with the collaborative economy largely based on platforms of exchange and provision, both in grassroots efforts and in business ventures.

The success of collaborative economy business platforms coincides with the fluidity between production and consumption. Where contemporary consumers are increasingly taking over work in their consumption practices through self-service and prosumption (Bauer and Gegenhuber 2015; Gabriel and Lang 2015; Ritzer 2014; Shirky 2008) and work practices are increasingly framed in consumption terms of fun or leisure (Jenkins and Delbridge 2014; Boje and Baskin 2011), the differentiation between work and consumption is seemingly thinning. Consequently, consumers offer their input and resources to organizational processes of production, while consumers' tangible and intangible resources, such as their possessions and skills, are lucrative and increasingly accessible on collaborative economy platforms.

The ability to blur the boundaries between consumption and work opens up new possibilities for value appropriation by businesses, who can profit from the aforementioned privately owned, micro-level capital made available on their platforms. Thus, as Astra Taylor (2014, p. 29) discusses, the 'communal spirit and capitalist spunk, play and work, production and consumption, making and marketing, editorializing and advertising, participation and publicity, the commons and commerce' become indistinguishable. On the other hand, forms of worker representation are challenged by the changing definitions, conditions, and sites of work: what counts as work, and who engages in work, become more difficult to pin down. Such blurring makes organizing and representing workers in the collaborative economy fraught with difficulties. The influence of collaborative economy business platforms on users as consumers and producers is an important topic in current discussions. Yet, users' agency to influence their condition and forms of collective organization and representation have received less attention.

In this chapter, we address the dearth of research on the means and forms of collective representation in collaborative economy business models. We focus on the ability of collaborative economy platforms to harness the working consumers' resources and their willingness to engage in value-producing activities that they do not primarily consider work. Such activities could be described as exploitative to the extent that they create profit through work without sharing profits or reimbursing the labour spend. While such activities are central to the collaborative economy at present, the role of exploitation as a fundamental driver of value in the collaborative economy context is not yet well understood – there is a perception of positive potentiality toward value sharing and liberation in the collaborative economy (e.g. Balaram 2016; Chase 2015), which is different compared to discussion on digital labour and platform capitalism (e.g. Scholz 2013). For instance, Gabriel and Lang (2015) fail to mention unpaid labour in the added category of consumer as worker in the updated treatise of the various images of the consumer. In short, traditional perspectives on work and consumption struggle to make sense of the new categories brought about by the collaborative economy.

In this chapter, we approach the question: *What are the possibilities and forms of collective representation in the collaborative economy?* Using a synthesizing approach to current literatures on collaborative consumption, platform-based collaborative economy and the few concurrent examples of collective action for justice and rights, we illustrate the struggle between individual and collective interests, and theorize on the potential of any collective representation to emerge. To date, organized collective action has been omitted from much of the discussion around collaborative consumption and the collaborative economy, while participants and stakeholders of the collaborative economy are seen to fight back from time to time (e.g. the mass demonstrations around working conditions at Uber or the impact of AirBnB on host cities). Yet, the emergence of new economic forms, the entanglement of production and consumption, and the possibility of exploitation within the unregulated grey area of platform-based collaboration complicate previous models of worker representation. Our literature review in this emergent context of collective representation in the collaborative economy outlines directions for conceptual development in this context guided by an underlying theory of injustice (Webster and Watson 2002).

To begin, we illustrate the origins of the collaborative economy as an emancipatory project and its subsequent development as a business model. We then review the roles of work and consumption among users of collaborative economy platforms. We discuss this in terms of prosumption – how the working consumer on collaborative economy platforms engages in both social and economic production. The relationship between the user and the platform, we argue, is often problematic and a cause of injustice for the participants as usership increases, but regulation remains opaque. How this injustice may lead to action, and particularly forms of collective representation, is treated in the final section of the chapter.

The business of collaborative consumption and collaborative economy

The radical idea of collaborative consumption in the collaborative economy is that through collaborative interactions – such as sharing – communities can collectively satisfy their members needs and develop each other's skills. Sharing dormant assets lessens the reliance on ownership and need for consumption and can thereby promote social and environmental sustainability. The early works of the collaborative economy evangelists (particularly Botsman and Rodgers 2010) popularised the interest in these new forms of economy. Initial ideas proliferated and produced various portrayals and neologisms around this central phenomenon/practice. These represented different but interrelated domains: *sharing economy, access economy* and *peer economy* for sharing and giving access to resources to peers in a community; *on-demand economy* and *gig economy* for performing task-based, temporary work on digital platforms; *grass-roots economy, local economy* and *alternative economy* for building localised and politicised alternative models; and *circular economy* supporting recycling and reuse. Collectively, these phenomena materialise in the foundational areas of everyday life, namely, housing, sustenance, travel, finance, and employment. They further connect previously distinct categories of the economy of production, consumption, market, and household in new dynamic patterns.

In this chapter, we use of the concept of collaborative economy, based on two considerations. First, the collaborative economy exemplifies a 'socio-economic order and technical systems that connect to historical, cultural, and technological developments in society and economic collaboration' (Wahlen and Laamanen 2017, p. 117). Here, collaborative economy stands for a seamless connection of network technologies with large economic potential, and social and environmental promise while drawing from and revitalising a community-oriented, participatory economic discourse and practice. Second, collaborative economy is increasingly becoming the concept utilised in public policy discourse in relation to various venture capital-based commercial applications. For instance, the recent report of the European Commission (2016b, p. 5) defines the collaborative economy as consisting of

> business models where activities are facilitated by collaborative platforms that create an open marketplace for the temporary usage of goods or services often provided by private individuals. The collaborative economy involves three categories of actors: (i) service providers who share assets, resources, time and/or skills – these can be private individuals offering services on an occasional basis ('peers') or service providers acting in their professional capacity ('professional services providers'); (ii) users of these;

> and (iii) intermediaries that connect – via an online platform – providers with users and that facilitate transactions between them ('collaborative platforms'). Collaborative economy transactions generally do not involve a change of ownership and can be carried out for profit or not-for-profit.

The definition illustrates how various actors come together in the collaborative economy. Defining the collaborative economy as above illustrates the attempt on part of governmental agents (here the EU) to make sense of and make actionable a contemporary conduct that has hitherto proven baffling to policy makers and public administration. The above definition restricts collaborative economy to business models and indeed it is possible to differentiate commercial ventures from local solidarity systems, virtual platforms from physical exchange networks, for-profit and non-profit (including public administration and community-based) initiatives, and their hybrids. The collaborative economy is arguably more than (few) business models and its functioning more complex than the definition alludes to (particularly regarding service providers and users). As our focus here is to examine commercial platforms, in the following we dissect the relation between platforms and their users.

Generally, research on collaborative consumption and the collaborative economy highlights the emancipatory and beneficial impact of sharing to individuals and communities, both online and offline, while noting the ambivalence of sharing schemes (Aigren 2012; Botsman and Rogers 2010; Balaram 2016; Chase 2015; Slee 2015; Wahlen and Laamanen 2017). The roots of collaborative consumption were in community building through principles of commons, trust, reputation, and reciprocity (Botsman and Rogers 2010). The assumption is that collaborative practices and reciprocities build mutual appreciation of participants engaging in economic collaboration (Bala and Schuldzinski 2016). Reciprocity affords symmetry in exchange while regulating expectations and outcomes (Polanyi 2001). However, it is a relationship that neither dissolves power between participants or groups nor their inclination to use it. According to Sahlins (1972), reciprocities can range from generalised, altruistic forms with a non-linear and indefinite character to negative reciprocities that seek maximum gain at the expense of others. Between these extremes lies balanced reciprocity that illustrates a tit for tat, direct and temporally-bound relationship.

From the perspective of the participation and provision of capital in reciprocal networks, local collaboration and opportunities of engaging in alternative, sustainable forms of work and consumption have benefits to the participants. However, there are shadows in paradise. New forms of economic collaboration have involved revamping capitalist modes of value extraction, using the moniker of "collaboration" to appear empowering and reciprocal. Yet, the centrality of digital platforms in connecting users – here both the occasional and professional providers and recipients of service – retains power and value with the owners and operators of the platforms, i.e. the venture capitalist and commercial entities. Critics have denounced how the new business models 'take advantage of two-sided markets and the platform-based economy, particularly those involving collaboration or sharing' in which a 'winner takes all' model dominates (Valenduc and Vendramin 2016, p. 8; see also Taylor 2014).

These commercial entities, acting as middlemen, are central to collaborative platforms, although their removal was one of the promises of the collaborative economy. Indeed, the potential benefits of community sharing must be weighed against the perils of exploitation: the 'new empowered and collaborative ways of consuming in the sharing economy, have become accused of consumer/worker exploitation and tax avoidance' suggesting 'that CC [collaborative consumption] and SE [sharing economy] are nothing but the emperor's new clothes, capitalism amplified and crowdsourced' (Wahlen and Laamanen 2017, p. 124). The ambivalence of such models between collaboration and profit-making can be further highlighted with 'sharing as non-commercial, person-to-person, social interaction … motivated by generosity … [while] economy suggests market transactions – the self-interested exchange of money for goods or services' (Slee 2015, p. 11). A potential intensification of exploitation of the participants, both consumers and workers, is a central problem with and critique of the collaborative economy (see e.g. Gabriel and Lang 2015; Slee 2015; also Zwick et al. 2008).

Relating reciprocity to the current collaborative economy, Evgeny Morozov (2013) questions whether sharing is a "feel-good utopia" driven by an altruism that is fully compliant with market logic (also Codagnone, Biagi and Abadie 2016). For Trebor Scholz (2016, p. 3) this precise logic turns people and their possessions into assets exemplifying 'the financialization of the everyday 3.0' (see also Codagnone, Abadie and Biagi 2016; Martin 2002; Slee 2015). The success of this logic is based on masking the precariousness of employment and wages, increasing inequality, and undermined rights of participants and social environments through customer experience and convenience. In this economy task-based, temporary work is the norm, not the option, and earning a little extra money on dormant resources can become exploitative rather than liberating. The platform is both shopping mall and factory (Scholz 2013). Rather than the humanistic notion of together in collaboration, participants are vetted against each other. How this has come to be, is what we illustrate next.

The consumer as producer and producer as consumer

Production and consumption are historically perceived as diametric economic activities, one for the creation, the other for assimilation of goods and services. Consequently, this division frames actors in two distinct economic roles: the consumer and the producer. Endrissat, Islam and Noppeney (2015) have argued that, historically, the relative position of these two roles has been variable, with contemporary work moving toward a fusion of consumption and production. Collaborative consumption is emblematic of this increasing intertwinement, with connective digital technologies providing tools that further confound these traditional roles. Particularly, the concept of prosumer (Toffler 1980; Cova and Cova 2012; Ritzer 2014; Ritzer and Jurgeson 2010) establishes the dual role of the economic actor as both the provider and the beneficiary who subsequently become 'agents of their own destinies' (Cova and Cova 2012, p. 149). Prosumers might be described as 'the person whose technology-mediated ingenuities would reach new heights of creativity and self-sufficiency, overcoming alienation and forging a better world' (Comor 2011, p. 310). Such creativity emerges in co-creation, which empowers community participants while simultaneously challenging the production-consumption dichotomy. Bonsu and Darmody (2008, p. 356) illustrate how engagement offers '...exchange and use values to the consumer in its contribution to consumer self-fulfilment through participation in affective community networks and its sociopolitical engagement based on respect and mutual sharing'.

Yet the thin line between liberation and exploitation becomes evident when prosumption and co-creation takes place on commercial platforms. Collaborative technological platforms (and the Internet in general) were initially seen as carrying the promise of equal participation in various social, political and economic activities. These were, as the retelling of the origins of the Peers community or AirBnB illustrate, perceived as "social movements" themselves (Balaram 2016; Morozov 2013; Slee 2015). However, there is a disconnect between the ownership of the intermediary commercial system by corporations and social exchange between users. While platforms provide increased choice, possibilities for interaction and utilisation of dormant resources for users, safety and workers' rights may become compromised.

The work by consumers as content creators relates to how consumers perform various forms of labour for businesses. This consumer work can be consensual, in cahoots with commercial organisations, to develop products and provide services, but also build social identity and communal relationships, which may provide opportunities for profit-making (e.g. Ross 2013). As a result of these co-creation processes, the consumer as producer may be exploited for their effort (Bauer and

Gegenhuber 2015; Bonsu and Darmody 2008; Cova and Dalli 2009; Zwick, Bonsu and Darmody 2010). Problematically, the consumer-worker is not remunerated for their input – 'know-how, enthusiasm and social cooperation' – but 'typically pay what the marketing profession calls a "price premium" for the fruits of their labour, as the use value provided by co-created commodities is said to be higher than that which can be achieved through standardized production's rationalized systems' (Cova and Dalli 2009, p. 327).

Co-operatively, any value co-created in the collaborative economy can be contained in and redistributed to the benefit of the community. On commercial platforms, co-created value is allocated to the platform owners and hence their centrality to and authority over resource allocation is not altered. Ritzer and Jurgensson (2010) claim that any capitalist control over the prosumer will prove difficult, as prosumers can resist exploitation by withdrawing their labour. Moreover, they argue that as abundance is produced through both free service and free labour, rationality-based calculation or exploitation become irrelevant concepts. Nonetheless, digitalization masks the productive nature of engagements of individual participants. As such, business models in the collaborative economy easily appropriate the value of individuals' engagement through various mechanisms. Others have illustrated how a mask of benevolence and fun hides the coercive features of social media technologies. By framing activities as entertainment and experience (e.g. "playbour"), their exploitative nature becomes effectively blurred. This leads to free (consumer) work and collaborative production, which particularly on digital platforms is amply available as networks grow, and participants create content (e.g. Beverungen et al. 2015; Fuchs 2014; Slee 2015; Taylor 2014; Zwick and Bradshaw 2016). As such, work and its exploitation extends beyond traditional contractual labour and physical work to life and its practices in general (Fleming 2014; also Bauer and Gegenhuber 2015; Huws 2016).

A platform should not be thought of, therefore, as a disinterested space of community participation. Scholz (2016, p. 2) assigns some of the blame to the users: 'your friendly convenience is, for many workers, a low-wage, precarious trap. But you, on the other hand, can listen to your very own Spotify account in an Uber taxi … You are promoted to middle management, entitled to fire your driver'. The semblance of agency and choice on the side of service consumption is paralleled by an increasing submission to demands on the service provision side. The idea that the collaborative economy platform may lead to servitude has led some writers to imagine a dystopian collaborative economy future, where 'a *small* number of people something like neofeudal masters, lords with a corncucopia [sic] of on-demand just-in-time luxury serv*ices* at their fingertips … by making a very *large* number of people glorified neo-serv*ants*…butlers, maids, chauffeurs, waiters, etcetera' (Haque 2015, emphasis original).

The relations between platforms and service providing prosumers is a cause of friction. The agentic powers of the emancipated prosumer are contained within the platform since platforms retain the power to exclude a user if they do not perform according to their contract. Griswold (2016) documents how angered Uber drivers demonstrate against the unilateral decision of the ridesharing platform to boost demand in slow periods by reducing fares. Drivers call the platform a dictatorship, unresponsive to the needs or grievances of its service-providing users. The case of Uber clearly demonstrates how the platform can be seen as a locus of power, generating discontent and possible action by workers. The point when injustice and grievance become mobilizing is what we look into next.

Injustice in the collaborative economy

Given the issues of power imbalance, negative reciprocities and exploitation in the collaborative economy, when do these become large enough to incite action? Snow and Soule (2010) discuss the centrality of grievance and perception of injustice for any corrective action to emerge. A grievance only becomes mobilizing when it is shared amongst a number of participants and felt substantial enough to warrant action. There are various ways in which unfairness and injustice can be examined to elaborate why (or why not) grievances become experienced and expressed individually and collectively. In what follows, we use Nancy Fraser's (2005, 2009) theory on social justice to understand the nature of a mobilizing grievance.

According to Fraser, justice incorporates the cultural dimension of recognition, the political dimension of representation and the economic dimension of distribution to achieve parity of participation: 'justice requires social arrangements that permit all to participate as peers in social life' (Fraser 2005, p. 73). Cultural injustices of misrecognition and status inequality ensue from hierarchical patterns of cultural value between social groups, whereas unequal voice and access in decision-making cause fundamental political injustice. Political injustice is consequently chiefly concerned with representation that problematizes governance structures and decision-making procedures and particularly individuals' access to participation on par. Economic injustices arise through exploitation, marginalisation, and deprivation of particular social groups in the structures of society. Collectively these forms of injustice have in common the limitation of the participative parity of participants, in which members are not able to act as equal members of the community or society.

We suggest that to resolve economic and political injustices a certain group of actors have to be recognized as such victims of injustice. Following Fraser's model, the case of prosumers suggests problems of non-recognition and cultural domination. Non-recognition suggests efforts to keep prosumers and their plight, through specific discursive practices, invisible to society in general, and particularly to places of political representation. One of the traditional ways of accomplishing that feat is to misrepresent and undervalue the actual representativeness and importance of the phenomena. Therefore, one important task is to demonstrate the pervasiveness of prosumers and their fight against exploitation in the collaborative economy.

Cultural domination suggests that prosumers are subjected to forms of interpretation that are alien to their reality. Therefore, one essential step is to recognize them as a specific group with their own particular grievances that demand a different analytical framework. Specifically, the pervasive tendency to view collaborative economy actors either in wholly communitarian terms, or when seen as economic actors, to frame them as "entrepreneurs". Both of these designations fail to recognize actors as workers and producers, and staves off their mobilization via various forms of worker representation.

Political domination, on the other hand, refuses an equal voice to participants in deciding the course of society. While collaborative economy platforms often have built into their digital systems elaborate ranking and rating systems, these tend to be heavily biased toward consumer satisfaction and provide less space for workers' concerns about the conditions of work or the constitution of the platform. Further, such ranking systems tend to be highly individualistic and rarely provide channels for collective participation or mobilization.

In terms of economic domination, several aspects of collaborative economy platforms may marginalize economically weaker parties, and diminish their ability to fully participate in the construction of the collaborative economy. First, algorithmic controls that highlight in search queries the most popular, best-selling or most "liked" products and services may lead to a "winner-takes-all" economy which systematically marginalizes participants and amplifies initial inequality conditions. Second, collaborative economy systems depend on assets and property already held by individuals, and facilitate the drawing of rents on that property. The dependence on initial property endowments means that the redistributive function of such platforms is likely to be overshadowed by the funnelling of resources to those already possessing the initial conditions for extracting rents, excluding those who lack the initial property.

Taken together, the cultural, political and economic aspects of collaborative platforms contribute to reducing participative parity so as to diminish the emancipatory potential of these platforms. The struggle for political participation and against economic exploitation has been too long linked to traditional identities of physical and contractual workers, and has not kept up with the new configurations of capital and labour represented in the collaborative economy. As such, current theories of collective representation help to keep invisible the particular struggles and abusive conditions that are experienced by this new category of hybrid economic actors. Political and economic action against injustice demands then an effort in theorizing the specificity of this new hybrid identity and their collective action.

In what follows, we aim to render visible and offer a first theorization of various ways in which collective action has taken place in the collaborative economy. It takes diverse forms, from discreet mobilisations and self-organization of disgruntled workers, organizing efforts stemming from traditional organized labour, to suggested models of shared governance and ownership. Together these examples illustrate the potential and problems of representation in the collaborative economy (cf. Huws 2016) as well as some approaches to relieve injustice.

Possibilities for collective representation in the collaborative economy

Building solidarities and reciprocities in the collaborative economy can prove to be challenging. Challenges relate, first, to the relationship between those employed in the traditional economy and those in collaborative economy, who, in the framework of the traditional capitalism, are competing for scarce resources. Secondly, it relates to the solidarities between platform prosumers. Is there a way in which servitude of the providing user, masked by the experience of the beneficiary user, could be subjected to critique? Reciprocities and solidarities are central to any collective action. In the traditional capitalist model, worker and consumer interest can come together locally with the threat of adverse community impact (cf. Fantasia 1988). Is there a possibility of generating joint identification for collaborative economy prosumers that would support collective resistance to platform power? We assume that common identification can enable representation and more radical politics. There are different affordances to how actors build reciprocities, solidarities and joint structures, which we discuss in terms of *discrete mobilizations, traditional labour organising, shared governance* and *common ownership.* The affordances may explain why resistance does not seem to be sustainable or leading to enduring change in the collaborative economy context. We will illustrate these in the following.

Classically regulatory actions regarding corporate conduct and general protection of workers and consumers has been the outcome of various consumer movements and the labour movement. Demands for fairer practices, inclusion, and voice have been more or less politically radical. The approaches range from attempts to transform capitalism through cooperatives and mutualist societies, whereby workers and consumer aim to regain their independent power, to less radical approaches ameliorating the effects of capitalism with increased information and participation in the conduct of business via government agencies providing consumer protection and information, and collective bargaining and co-determination models, such as works councils (Gabriel and Lang 2015; Hilton 2009; Moraes, Szmigin and Carrigan 2010; Silver 2003).

The concerns of consumers relating to economic responsibility and fairness, welfare of communities and the environment and towards global, systemic risks have facilitated the emergence of alternative models of exchange, including the rise of the collaborative economy, which taps into the resilient concerns and values of market participants (Moraes, Szmigin and Carrigan 2010; Wahlen and Laamanen 2015). Still, illustrated in connection with commercial platforms of the collaborative economy, the interests of platforms and their users are highly volatile. The dynamics of platform relationships are embedded in lacking wage contract, gig-based employment, and peer-to-peer exchanges, so that representation of interests on a collective level may not follow the traditional models of industrial society. Moreover, commercial collaborative economy communities are pseudo-communities, in the sense that the social bonds connecting members may be short-lived, and not rooted in deep or durable social or institutional forms. In the collaborative economy, questions of ownership, rewards and responsibility are difficult to resolve. This difficulty reflects situations where capital is owned by participant users, yet managed by the platform, which extracts a (high) rent, but often assumes none of the responsibility, and often without remuneration for the participant users.

Similarly, the motivations of participants can be as varied as the business models and platforms where the activities take place. This means that not all users care what the conditions and outcomes of their service production are, nor do they necessarily want to reciprocate or create value for the larger community. Thus, beyond its promise, collaborative economy may also contribute to the creation of atomised, hypercapitalist collaboration beyond traditional forms of regulation (Slee 2015; Wahlen and Laamanen 2017). Such developments are apparent in the various approaches taken to global players (such as AirBnB and Uber) in various locations and the rather lethargic supranational and national governance (cf. European Commission 2016a, b; U. S. Department of Commerce 2016).

Despite all these limitations, it is possible to identify some collaborative economy models that address the grievances and problems. These take diverse forms, from discrete mobilizations and self-organization of disgruntled workers, through organizing efforts stemming from traditional organized labour, to shared governance and ownership. In *discrete mobilizations,* workers attempt to self-organise on a local level or in an industry, often around and in demonstrations, such as those of delivery workers of Foodora and Deliveroo, or Uber drivers. Here the aim is to address and make visible the immediate grievances of the users, who are generally seen as engaging in work on the platform for fun (e.g. Ciccarelli 2016). Local demonstrations may also take aim at the social impact of collaborative economy platforms and their practices. In various cities around the world, taxi drivers have demonstrated against the threat to their industry by "unregulated" platforms. Prosumers online can similarly make claims over their input, such as the Wages for Facebook campaign, drawing on the feminist ethos of the International Wages for Housework campaign, demonstrates.

Some discrete mobilizations can lead to action by authorities. This has been the case with decisions to address the actions of platforms. Propositions to restrict activities of platforms and their users, such as AirBnB, on the basis of social impact have been resolved differently in various locations. Short-term rentals are becoming more strictly regulated in Berlin and London whereas a proposition on restrictions was defeated in San Francisco.

The traditional categorical distinction between consumption and production as separate activities is increasingly disappearing. With this, the argument could be made that representational structures of the traditional economy or the governance of conduct are similarly ill-fitting. Representation of consumer and worker interests as the less powerful participants in the marketplace has traditionally taken the form of regulation with clear areas of application. Hence, societal and economic peace was achieved through separation of production and consumption (in the Fordist deal; see Gabriel and Lang 2015; Lichtenstein 2013). In the collaborative economy, this separation is less meaningful. Still, some *traditional labour organising* makes efforts to combat the precariousness of employment (e.g. Balaram 2016). In general, attempts to unionise non-traditional settings with disconnected workers and/or high turnover (e.g. hospitality or fast food industries) have proved to be challenging to traditional labour movement actors. While some progress has been made by "alt-labor" workers' groups (e.g. Eidelson 2013; Lakhani and Tapia 2014), union-like activity in the collaborative economy is hindered by the atomisation of both the tasks and the workforce, members of which often work without an explicit employment contract as freelancers. New labour organizations, such as The Freelancers Union (FU), claim to represent freelance employees; yet critics claim that the FU cannot claim that it collectively represents the gig-workers and should be seen as nothing more than an insurance brokerage (see e.g. Paul 2014). Moreover, the globally dispersed, unconnected users performing discrete exchanges represent a form of neo-Taylorism that makes traditionally effective forms of collective organising difficult and accentuates the cost of organizing. Specific communities, such as the Turkopticon (https://turkopticon.ucsd.edu), aim to offer help to other Mechanical Turk users through reporting problems and employment experiences. Unions seem powerless, as with the declining influence and resources, the classical labour movement's support for organising in the collaborative economy remains weak.

Next, *shared governance* implies models of shared regulation where participant actors share a common set of rights and responsibilities in their exchanges of private, shared capital. This represents a more participatory model, rather than reactive form of current regulation. The shared forms of regulation are attempts to ensure active inclusion of participants as well as other stakeholder groups, and to disperse the power to influence the workings of the collaborative economy: '*"shared regulation" entails the redistribution of regulatory responsibility to parties other than government;* however, it differs from self-regulation because *businesses are only one of many parties involved*' (Balaram 2016, p. 38, emphasis original). Shared governance extends the traditional capitalist model by attempting to ameliorate problems of exclusion and exploitation. Addressing social injustice requires 'dismantling institutionalized obstacles that prevent some people from participating on a par with others, as full partners in social interaction' (Fraser 2005, p. 73). The collaborative economy can promise sustainable economic opportunity, yet this requires decentralised power in redistribution and equal access to resources (Martin 2016). If, however, platforms are structured to colonize and commodify, participation and co-creation becomes 'a veil of … empowerment in a world in which market power, in large measure, still resides in capital' (Bonsu and Darmody 2008, p. 366).

The shared governance perspective can be seen to be connected to the argument for shared value, which states that business profitability over the long term rests on a stakeholder-inclusive outlook, with economic value created in a way that concurrently addresses social needs and challenges, and from a societal perspective. The social organisation generating value is inconsequential, as the important factor is the benefit to cost ratio in producing the value (Porter and Kramer 2011). To manage and encourage collective value creation raises the key issue of individual motives for co-operating across interaction networks. From an organisational perspective, collective value creation is the interplay between an organisation's motivational system and its participants' mix of motives (cf. Bridoux et al. 2011).

Social justice further concerns the possibilities of representation in local and global contexts. The alter-globalisation movement has, amongst others, addressed questions of fairness and justice in global supply chains as well as the general problems of democracy in a world of global, connected capital and local, disconnected labour. Scaling up in the global setting becomes difficult as 'representation functions on the assumption that a certain portion of the constituency will have identical interests and creates a circumstance in which only these *shared* interests get expressed, but these shared interests become fewer and farther between as communities become more diverse and the scale of decision-making more global' (Maechelbergh 2014, p. 357). More active regulation from authorities could alleviate the more general grievances in the collaborative economy.

On collaborative economy platforms, there are few strong mechanisms for collective voice. The utopian discourse of the collaborative economy largely silences dissenting voices (Morozov 2013). For instance, the EU report on the collaborative economy is said to echo the platforms' opinions (Kowalsky 2016). A key political issue for the collaborative economy is the difficulty of regulating the platforms, which the EU reports' lethargic approach illustrates. The EU report (2016a, b) elaborates that there is no applicable legal structure to tackle issues in the collaborative economy, but these should be addressed on a case by case basis. Thus, the possibility of joint Pan-European governance and coordination is left an open question. An interesting case involving share regulation is the recently settled court case of Uber's identity as a transportation company. This case illustrates the problems of shared regulation, which on a larger scale rests on the various, diverging interests of participants. Authorities' perceptions of the industry that the platform operates are divergent, illustrating particular European dilemmas (Bershidsky 2016) that attempt to balance, on the one hand, differing perceptions and approaches to regulation in different member countries, and on the other, the perception of the positive impact of worker (and consumer) protection and extant labour market models.

Finally, if indeed workers in the framework of the traditional capitalism are competing for scarce resources regardless of their location in the traditional or the collaborative economy, ownership could be the most powerful alternative for addressing injustice. *Common ownership* is an attempt to radically change the current form, drawing from older models of socialism and the collaborative form (Parker, Fournier and Reedy 2007). Here, the different forms of capital are governed collectively in joint ownership, right to use and reciprocation; extending to recent developments in the non-commercial collaborative economy, such as solidarity economy and commons as well as platform collaborativism. Drawing on the ideas of shared governance, the bringing together of various stakeholders illustrates the alternative politics of the everyday, that of local action where economic activity involves 'producers, workers, consumers and citizens acting collectively and in solidarity' (Utting 2015, p. 1–2). Accounting for the intertwined roles of consumer and worker functions in new forms of cooperativism requires a recognition of the pervasiveness of prosumerism, the prosumer identity and their fight against cultural, economic and political injustice. Solidarity economy commons rest on building communities that jointly create and share competence, resources and sustenance, with or without common ownership (see e.g. Ostrom 2010). With platform-fostered collaborativism (Scholz 2016, 2017), the cooperative jointly owned form eliminates managerial control, and participant value co-creators have equal representation and rights in decision-making as well as access to the fruits of their work. Moreover, the model here is connected to securing jobs, not profits. Examples here are the car sharing co-op Modo and the co-operative online marketplace Fairmondo. Shared ownership can function as a mechanism to fend off the quasi-communality and increasing individualization built in to collaborative economy platforms and bring about a genuinely new and just economy.

Conclusion

The atomisation of provider-consumer connections leads to strengthening of the platform, while the middleman who was supposed to disappear with these new models remains central, if less visible. With this in mind, the question is what form does representation of the collaborative economy participants' interest against the corporate platform providers take. We argue that when individuals engage in value-creating activities on collaborative economy platforms, they exert little influence over the terms and conditions of this arrangement, both as workers and consumers. With regard to economic injustices, platforms appropriate content, shun responsibility, and externalise welfare. Indeed, individuals in collaborative economy contexts offering their input are independent contractors without access to the benefits of traditional employment relations (e.g. Slee 2015). This economic injustice clearly is an issue for redress, when collaborative economy platforms make the rules while carrying little responsibility. Asymmetries of benefits and responsibilities illustrate how the moral responsibilities and externalities to societal stakeholders are generally neglected.

We suggest that, at the root of the problems of economic exploitation and political misrepresentations, rests a fundamental misrecognition of the nature of prosumers and their specific challenges. The cultural conditions that frame prosumers on platforms as entrepreneurs, freelancers and independent contractors simultaneously naturalize the extension of the economic into the everyday lives of people and their possessions. Consequently, traditional collective representation models are still mostly paralyzed on outdated distinctions between producers and consumers as economic actors, blind to the new category of prosumers as economic actors. The lack of understanding of the economic and political constraints experienced by the new generation of members of the new economy, of which the sharing movement is an essential part, limits our (and their) understanding and solidarity with their plight.

We illustrated several contemporary forms of collective organizing that could potentially be applied in the collaborative economy. Discrete mobilizations may be able to dislodge particularly evident injustices, once groups of users collectively perceive injustice. Traditional labour organizing may be feasible in some instances, although as argued above, several aspects of collaborative economies mitigate against traditional forms of labour organizing. Shared governance may allow innovations in the functioning of collaborative platforms that are more equitable and recognize the plurality of actors involved. Finally, common ownership may address, at their roots, the property regimes that allow individual extraction of wealth at the expense of community welfare. While each of these solutions requires further development, some combination of them, tailored to the particular platform contexts, may point to ways of achieving participative parity in the collaborative economy.

The implications of the collaborative economy for reframing the work and non-work spheres are both ideological and material, with ambivalent outcomes. Joining work and consumption through forms of community could be a form of political activism. Finding new ways to join these spheres might reflect ongoing anxieties about this split and distinguish between different forms of political activism and/or ideological reframing. Ideally, collective representation in the collaborative economy could build solidarities and reciprocities based on common experiences of digitalised, platform-based work: helping others by 'taking time to respond to a person one has never met, purely because one feels an empathy with their predicament, is an aspect of this new form of solidarity that has thrived in the Internet age' (Gabriel and Lang 2015, p. 223; also Taylor 2014). Similarly, the implications of the collaborative economy for reframing what is work and consumption, and private and public spheres, are currently managed by the platforms, with ambivalent outcomes for participants. Finding new ways to join these spheres is the key to mobilizing in the collaborative economy.

References

Aigren, P. (2012). *Sharing: Culture and Economy in the Internet Age*. Amsterdam, NL: Amsterdam University Press.

Bala, C., & Schuldzinski, W. (2016). Einleitung: Neuer sozialer Konsum? In C. Bala & W. Schuldzinski (eds.), *Prosuming und Sharing – neuer sozialer Konsum. Aspekte kollaborativer Formen von Konsumtion und Produktion* (pp. 7–29). Düsseldorf, DE: Verbraucherzentrale NRW.

Balaram, B. (2016). *Fair Share: Reclaiming power in the sharing economy* [RSA publication, January 2016]. Retrieved from https://www.thersa.org/discover/publications-and-articles/reports/fair-share-reclaiming-power-in-the-sharing-economy/#

Bauer, R.M., & Gegenhuber, T. (2015). Crowdsourcing: Global search and the twisted roles of consumers and producers. *Organization*, 22(5), 661–681.

Bershidsky, L. (2016, November 30). EU Court Will Resolve Uber's Identity Crisis. *Bloomberg*. Retrieved from https://www.bloomberg.com/view/articles/2016-11-30/taxi-or-tech-eu-court-will-rule-on-uber-s-identity

Beverungen, A., Böhm, S., & Land, C. (2015). Free Labour, Social Media, Management: Challenging Marxist Organization Studies. *Organization Studies*, 36(4), 473–489. doi: 10.1177/0170840614561568

Boje, D.M., & Baskin, K. (2011). Our organizations were never disenchanted: Enchantment by design narratives vs enchantment by emergence. *Journal of Organizational Change Management*, 24, 411–26.

Bonsu, S.K., & Darmody, A. (2008). Co-creating Second Life: Market-Consumer Cooperation in Contemporary Economy. *Journal of Macromarketing*, 28(4), 355–368.

Botsman, R., & Rodgers, R. (2010). *What's Mine is Yours: How Collaborative Consumption is Changing the Way We Live*. London, UK: HarperCollins.

Bridoux, F., Coeurderoy, R., & Durand, R. (2011). Heterogeneous motives and the collective creation of value. *The Academy of Management Review*, 36(4), 711–730.

Chase, R. (2015). *Peers Inc : How People and Platforms Are Inventing the Collaborative Economy and Reinventing Capitalism*. New York, NY: PublicAffairs.

Ciccarelli, R. (2016, October 12). Delivery workers strike in latest flare up of gig economy conflict. *il manifesto*. Retrieved from http://ilmanifesto.global/delivery-workers-strike-in-latest-flare-up-of-gig-economy-conflict

Codagnone, C., Abadie, F., & Biagi, F. (2016). *The Future of Work in the 'Sharing Economy'. Market Efficiency and Equitable Opportunities or Unfair Precarisation?* [Institute for Prospective Technological Studies, JRC Science for Policy Report EUR 27913 EN]. Retrieved from http://publications.jrc.ec.europa.eu/repository/handle/JRC101280

Codagnone, C., Biagi, F., & Abadie, F. (2016). *The Passions and the Interests: Unpacking the 'Sharing Economy'* [Institute for Prospective Technological Studies, JRC Science for Policy Report EUR 27914 EN). Retrieved from http://publications.jrc.ec.europa.eu/repository/handle/JRC101279

Comor, E. (2011). Contextualizing and Critiquing the Fantastic Prosumer: Power, Alienation and Hegemony. *Critical Sociology*, 37(3), 309–327.

Eidelson, J. (2013, January 29). Alt-Labor. *The American Prospect*. Retrieved from http://prospect.org/article/alt-labor

Endrissat, N., Islam, G., & Noppeney, C. (2015). Enchanting Work: New Spirits of Service Work in an Organic Supermarket. *Organization Studies*, 36(11), 1555–1576.

European Commission (2016a). *A European agenda for the collaborative economy* [COM(2016) 356]. Retrieved from http://ec.europa.eu/DocsRoom/documents/16881

European Commission (2016b). *European agenda for the collaborative economy – supporting analysis* [SWD(2016) 184]. Retrieved from http://ec.europa.eu/DocsRoom/documents/16881

Fantasia, R. (1988). *Cultures of Solidarity*. Berkeley, CA: University of California Press.

Fleming, P. (2014). *Resisting Work: The Corporatization of Life and its Discontents*. Philadelphia, PA: Temple University Press.
Fraser, N. (1995). From redistribution to recognition? Dilemmas of justice in a 'post-socialist age. *New Left Review*, I/212 (July-August), 68–93.
Fraser, N. (2005). Reframing justice in a globalizing world. *New Left Review*, 36 (November-December), 69–88.
Fraser, N. (2009). Scales of Justice: Reimagining Political Space in a Globalizing World. New York, NY: Columbia University Press.
Fuchs, C. (2014). Digital prosumption labour on social media in the context of the capitalist regime of time. *Time & Society*, 23(1), 97–123.
Gabriel, Y., & Lang, T. (2015) *The Unmanageable Consumer*. (3rd ed.). London, UK: Sage.
Griswold, A. (2016, February 23). The scariest thing about the gig economy is how little we actually know about it. *Quartz*. Retrieved from http://qz.com/614218/the-scariest-thing-about-the-gig-economy-is-how-little-we-actually-know-about-it/
Haque, U. (2015, June 8). The Servitude Bubble. Tech isn't really making a "sharing" economy. So what is it making? *Medium*. Retrieved from https://medium.com/bad-words/the-servitude-bubble-c9e998c437c6#.8ugqx2t5g
Hilton, M. (2009). *Prosperity for All: Consumer Activism in an Era of Globalization*. Ithaca, NY: Cornell University Press.
Huws, U. (2016, June 1). Logged In. *Jacobin*. Retrieved from https://www.jacobinmag.com/2016/01/huws-sharing-economy-crowdsource-precarity-uber-workers
Jenkins, S., & Delbridge, R. (2013). In pursuit of happiness: A sociological examination of employee identifications amongst a "happy" call-centre workforce. *Organization*, 21(6), 867–887.
Kowalsky, W. (2016, June 13). What a wonderful new world: The sharing economy. *Social Europe*. Retrieved from https://www.socialeurope.eu/2016/06/wonderful-new-world-sharing-economy/
Lakhani, T., & Tapia, M. (2014). Organizing the fragmented unorganized. An examination and history of the fast food movement. *Perspectives on Work*, 18, 22–25 & 106.
Lichtenstein, N. (2002). *State of the union: A century of American labor*. Princeton, NJ: Princeton University Press.
Maeckelbergh, M. (2014). Social movements and global governance. In M. Parker, G. Cheney, V. Fournier & C. Land (eds.), *The Routledge Companion to Alternative Organization* (pp. 345–358). London, UK: Routledge.
Martin, C. J. (2016). The sharing economy: A pathway to sustainability or a nightmarish form of neoliberal capitalism? *Ecological Economics*, 121 (January 2016), 149–159.
Martin, R. (2002). *Financialization of Daily Life*. Philadelphia, PA: Temple University Press.
Moraes, C., Szmigin, I., & Carrigan, M. (2010). Livin production-engaged alternatives: An examination of new consumption communities. *Consumption Markets & Culture*, 13(3), 273–298.

Morozov, E. (2013, October 14). The 'sharing economy' undermines workers' rights. *The Financial Times*. Retrieved from https://next.ft.com/content/92c3021c-34c2-11e3-8148-00144feab7de

Ostrom, E. (2010). Beyond Markets and States: Polycentric Governance of Complex Economic Systems. *The American Economic Review,* 100(3), 641–672

Parker, M., Fournier, V., & Reedy, P. (2007). *The Dictionary of Alternatives*. London, UK: Zed Books.

Paul, A. (2014, October 28). A Union of One. *Jacobin*. Retrieved from https://www.jacobinmag.com/2014/10/freelancers-union

Polanyi, Karl. (2001/1947). *The great transformation: The political and economic origins of our time*. (2nd ed.). Boston: Beacon Press.

Porter, M., & Kramer, M.R. (2011). Creating Shared Value. *Harvard Business Review,* 89(1/2), 62–77.

Ritzer, G. (2014). Prosumption: Evolution, revolution, or eternal return of the same? *Journal of Consumer Culture,* 14(1), 3–24. doi: 10.1177/1469540513509641

Ritzer, G., & Jurgenson, N. (2010). Production, Consumption, Prosumption: The nature of capitalism in the age of the digital 'prosumer'. *Journal of Consumer Culture,* 10(1), 13–36.

Ross, A. (2013). In Search of the Lost Paycheck. In T. Scholz (ed.), *Digital Labor: The Internet as Playground and Factory* (pp. 13–32). New York, NY: Routledge.

Sahlins, M. (1972). *Stoneage economics*. Chicago: Aldine-Atherton.

Scholz, T. (Ed.). (2013). *Digital Labor: The Internet as Playground and Factory*. New York, NY: Routledge.

Scholz, T. (2016). Platform Cooperativism. Challenging the Corporate Sharing Economy. Retrieved from http://www.rosalux-nyc.org/platform-cooperativism-2/

Scholz, T. (2017). *Uberworked and Underpayed*. Cambridge: Polity.

Shirky, C. (2008). *Here Comes Everybody*. London: Penguin Books.

Silver, B.J. (2003). *Forces of labor: Workers' movements and globalization since 1870*. Cambridge, UK: Cambridge University Press.

Slee, T. (2015). *What's Yours is Mine. Against the Sharing Economy*. New York, NY: OR Books.

Snow, D.A., & Soule, S.A. (2010). *A Primer on Social Movements*. New York, NY: W.W. Norton & Company.

Taylor, A. (2014). *The People's Platform: Taking back Power and Culture in the Digital Age*. New York, NY: Metropolitan Books.

U.S. Department of Commerce (2016). *Digital Matching Firms: A New Definition in the "Sharing Economy" Space*. Retrieved from http://www.esa.doc.gov/sites/default/files/digital-matching-firms-new-definition-sharing-economy-space.pdf

Utting, P. (2015). Introduction: The challenge of scaling up social and solidarity economy. In P. Utting (Ed.), *Social and solidarity economy: Beyond the fringe* (pp. 1–37). London, UK: Zed Books.

Valenduc, G., & Vendramin, P. (2016). *Work in the digital economy: sorting the old from the new* (European Trade Union Institute Working Paper 2016.03). Brussels, BE: ETUI aisbl.

Wahlen, S., & Laamanen, M. (2015). Consumption, Lifestyle and Social Movements. *International Journal of Consumer Studies*, 39(5), 397–403.
Wahlen, S., & Laamanen, M. (2017). Collaborative Consumption and Sharing Economies. In M. Keller, B. Halkier, T.-A. Wilska & M. Truninger (eds.), *Routledge Handbook on Consumption* (pp. 116–130). Oxford, UK: Routledge.
Webster, J., & Watson, R. T. (2002). Analyzing the past to prepare for the future: Writing a literature review. *MIS Quarterly*, 26(2), xiii–xxiii.
Zwick, D., & Bradshaw, A. (2016). Biopolitical Marketing and Social Media Brand Communities. *Theory, Culture & Society*, 33(5), 91–115.
Zwick, D., Bonsu, S. K., & Darmody, A. (2008). Putting Consumers to Work: 'Co-creation' and new marketing govern-mentality. *Journal of Consumer Culture*, 8(2): 163–196.

"Foodsharing": Reflecting on individualized collective action in a collaborative consumption community organisation

Stefan Wahlen

Abstract

Food waste is increasingly seen as a social problem. The "foodsharing" collaborative consumption initiative is an attempt to mitigate this societal challenge. This chapter investigates the community structure and consumer culture associated with "foodsharing". As a collaborative consumption movement, food consumption is politicized in the "foodsharing" movement. With a growing community, the "foodsharing" community in Berlin (Germany) faced a conflict with the executive branch of consumer policy, the local food authorities. This chapter is in general interested in how the "foodsharing" community is set up and how the community organisation is combined with particular consumer identities. The chapter further investigates how politicization and a conflict with local food authorities lead to consumer activism against consumer policy. The chapter concludes with a theoretical reflection on collaborative consumption movements, highlighting how the boundaries between individual and collective action are continuously blurred and what consumer policy can learn from this contemporary empirical example.

Zusammenfassung

Lebensmittelverschwendung wird zunehmend als soziales Problem angesehen. Die Initiative „foodsharing" hat den Versuch gestartet dieses gesellschaftliche Problem zu lösen. Dieses Kapitel untersucht Organisationsstruktur und Konsumkultur welche mit der Initiative „foodsharing" verbunden sind. Als kollaborative Konsumbewegung wird der Nahrungsmittelkonsum in der „foodsharing"-Bewegung politisiert. Dieses Kapitel ist im Generellen an der Organisationsstruktur der Initiative „foodsharing" und damit verbundener Identität interessiert. Mit einer wachsenden Community und Bekanntheit geriet die „foodsharing"-Gemeinschaft in Berlin in einen Konflikt mit der lokalen

I. Cruz et al. (Hrsg.), *Contemporary Collaborative Consumption*,
Kritische Verbraucherforschung, https://doi.org/10.1007/978-3-658-21346-6_4

Lebensmittelbehörde. Daher wird im Speziellen untersucht inwiefern die Politisierung des Konsums im Rahmen des Konfliktes mit der lokalen Lebensmittelbehörde als Verbraucheraktivismus gegen die Verbraucherpolitik einzuordnen ist. Das Kapitel schließt mit einer theoretischen Reflexion über kollaborative Konsumbewegungen und zeigt auf, wie Grenzen zwischen individuellem und kollektivem Handeln kontinuierlich verschwimmen. Ferner wird reflektiert inwiefern Verbraucherpolitik aus diesem zeitgenössischen empirischen Beispiel weiter entwickelt werden kann.

Keywords

foodsharing, collective action, community organization

Introduction: collaborative consumption and food waste

Food waste is increasingly seen as a problem for the environment and therefore also as a societal challenge (Aschemann-Witzel et al. 2015, Wahlen and Winkel 2017). Political debates on food waste are embedded in discussions surrounding sustainable consumption and production, as for example in the United Nations (UN) sustainable development goals (United Nations 2016). Food is thrown away at different points in the food chain, from primary production in agriculture to manufacturing and trade, and finally by consumers and in private households. According to the Food and Agriculture Organization (FAO) of the United Nations, around 1.3 billion tons of edible food are being wasted each year around the world (FAO 2013).

The causes and consequences of food waste are many (Quested et al. 2013). Regarding food waste, consumer policy initiatives contribute to both the reduction and the spread of food waste. In France, for example, the national government has since 2015 admonished supermarket chains to donate excess food to charitable organizations and thus to combat food poverty. On the other hand, the best before date and regulations on standardized size and shape of fruit and vegetables contribute to food being disposed of, even though it is (still) edible and hence suitable for consumption. In many cases, there is an observable "fetishism of immaculacy": food which does not match the ethos of impeccability is disposed of rather than eaten. For example, the best before date, originally introduced by consumer policy for the protection of consumers, is often interpreted by consumers as being scientifically accurate fact, and as a result food exceeding that date tends to be thrown away rather than eaten. However, this should not be misinterpreted as ascribing the responsibility solely to the consumer. Regarding shape and size of fruits and vegetables, a vast amount of food is disposed of along the food supply chain even before entering the consumer market. This "fetishism of immaculacy" is increasingly having the effect that only products which correspond to a certain ideal come to market. Food that is still edible is becoming food waste at various moments in the food chain.

In the recent past, government, non-government organisations and consumers have all addressed the issue of food waste, which is seen as a societal challenge. Actors in the system of provision are involved in increasing debate on sustainable development and the question of how societies should develop in the future, and with it the day to day activity of individuals. Although there are numerous initiatives to reduce food waste, little has been achieved (Evans 2012; Aschemann-Witzel et al. 2015). This is where this chapter comes in: it is a case study of an initiative which highlights the tension between food waste, a collaborative consumption movement fighting that waste, and consumer policy.

The objective of this chapter is to approach food waste from a perspective of collaborative consumption (Wahlen and Laamanen 2017) and lifestyle politics (de Moor 2016). This chapter advances theoretical debates locating contemporary collaborative consumption in lifestyles, political action and social movements. It clarifies collaborative consumption as lifestyle politics by deciphering a complex nexus of community practices in the German "foodsharing" initiative. This makes it possible to disentangle the relationship between individual and collective action, emphasize the blurring of boundaries and reflect on the role of trust in a collaborative consumption community. The chapter proceeds as follows: the remainder of this introduction discusses the position of political consumption between individual and collective action. The rest of the introduction outlines an understanding of (collaborative) consumption as social practices and closes by clarifying the methodological approach. In the remainder of this chapter, the "foodsharing" initiative is considered as a case study of the blurring of the boundaries between lifestyles, politics and social movements. I would like to consider individual and collective action by scrutinizing the community structure as well as identity and the consumer culture associated with the "foodsharing" initiative. Finally, the results are discussed, with the aim of contributing to theories of contemporary collaborative consumption in conjunction with consumer policy and lifestyle movements.

Understanding consumption as social practices

In contemporary societies, we are witnessing increased marketization as well as commercialization. Consumers depend heavily on the market and on commercial contexts. Nevertheless, reducing consumption to market activity does not account for day to day activity that may be regarded as consumption (Warde 2016).

The dominant economic understanding of consumption as market exchange does not take household activity and non-monetary aspects of consumption into account and further promotes, along with its individualist growth-oriented argumentation, the destruction of livelihoods in environmental and social respects (Fridrich et al. 2014; Fridrich et al. 2016; Yates 2016). In order to overcome an individualist approach to consumption, this chapter offers an interpretation of consumption as social practices.

Consumption has to an increasing extent been studied from practice theoretical perspectives during the past decades. From this theoretical perspective, consumption in general, and eating as a moment of consumption in particular, is understood in a broader socio-cultural perspective as routinized conduct in everyday life (e. g. Warde 2016). An advantage of this approach is that consumption is not necessarily understood and reduced to individualized market activity, but sharing practices are seen as an everyday phenomenon in the living environment of households and families (Yates 2016; Niehof and Wahlen 2017). Reckwitz (2015) points out that theories of social practices are neither individualist nor collectivist. In a culturalist way of theorizing, consumers may be considered as agentic carriers in consumption practices, but they are not the locus of concern.

Contemporary collaborative consumption, considered as part of social practices, can be regarded as a process and in its theoretical aspects emphasises practical conduct in performance rather than focussing on mental activity (Warde 2016). In relation to food waste in a household context, it is suggested that eating practices are part of a nexus consisting of different activities such as planning, buying, storing, cooking, ingesting and disposing of food, in the sense of "doing food" (Winkel, Wahlen and Jensen 2016; Wahlen and Winkel 2017). Together, these procedural steps in practices make eating food possible. With this approach, practices form a nexus in which it is of particular interest to investigate the connections between the different activities which waste food – from modes of provision to food preparation and finally ingesting food.

In this regard, food waste and consumption are not only considered in the context of market withdrawal, but as a complex socio-cultural activity in everyday life. Between practitioners belonging to the same group, practices are subject to judgment with regard to their acceptance (Warde 2016, p. 41). Hence, there are socially acceptable modes or practices of provision. Moreover, in considering consumption as social practice, the ability to follow rules and instructions becomes evident. These rules pertain to the acceptability of the procedural steps of food provision, storage, and preparation.

Contemporary collaborative consumption practices thus relate to sharing activity in everyday life and household contexts (Niehof and Wahlen 2017; Wahlen and Laamanen 2017; Yates 2016). Understanding consumption as social practice has implications for consumer policy and the politics of consumption. Analysis of the "foodsharing" initiative as an alternative mode of provisioning food compared to regular market activity helps to shed light on the politics of practices and associated implications for consumer policy. In this vein, it is interesting to consider collectively shared modes of provision in practices individually enacted by carriers of the corresponding provisioning practices. Since practices are tied in with individual agentic carriers, conventionally speaking the consumer, practices always have a collective aspect. Hence, the question which arises is how communities can support social change in reducing food waste, to the extent that collectivity in practices is promoted. In that collectivity of social practices, trust plays an important role, because trust can be considered as a "key property of the collective organization of social relations" (Kjaernes, Harvey and Warde 2007, p. 8). The collective nature of practices in the "foodsharing" community may help us to reflect on individualized collective action and the organization of social relations.

Political consumption as individual and collective activity

Political consumption and lifestyle politics can be located at the intersection of governmental consumer policy and consumer political activism. On the one hand, political activities by governments and administrations endorse initiatives that aim to reduce food waste. On the other hand, consumers are politically active in so far as they are involved in social movements. Consumers in contemporary collaborative forms of consumption such as "foodsharing" are concerned with influencing the (political) system in order to achieve (social) change. In that sense collaborative forms of consumption might address issues on the political agenda (Wahlen and Laamanen 2015), working towards less food waste. Government policy and political consumption have in common that attempts are being made to perform consumption in a particular way, in the hope that social change may be brought about by influencing the development of living conditions (Dubuisson-Quellier 2015), for example in terms of sustainable development. Consumption practices in terms of political consumption and lifestyle politics are accordingly understood in a broad sense as forms of political participation (de Moor 2016). Baringhorst (2015) explains that everyday practices of individual consumers are charged in moral as well as political ways, leading to lifestyle politics. These political activities aim to achieve a certain political objective, in particular here to reduce food waste. Accordingly, political activity is not necessarily understood in the traditional sense as a social movement and street protests, but rather in privately situated activities aimed at changing the living environment (Dubuisson-Quellier 2015). Yet again, however, it is questionable to what extent a fragmentation of everyday life of individualized activity is also a questioning of social and political power structures (Baringhorst 2015).

Lamla (2015) raises the question of how the politicization of the consumer market sphere can be understood as social movement activity and whether such activities weaken or strengthen current democracy. In a similar but a somewhat narrower vein, Stolle and Micheletti (2013) see the political nature of products and political consumerism as voting with the shopping basket in order to change institutional or market practices. In their understanding of political consumerism as a move towards more ethical product acquisition, they bring the idea of individualized collective action to the fore. As such, consumption activity is understood as citizen engagement concatenating individual self-interest (consumer choice) and responsibilities for the general good (citizen virtues). They define such individualized collective action as "responsibility-taking for common well-being through the creation of concrete, everyday arenas on the part of the citizens alone or together with others to deal with problems that they believe are affecting what they identify as the good life." (Stolle and Micheletti 2013, p. 26)

In line with the understanding of consumption as social practice, individualist collective action oscillates between individual and collective levels of social activity. In this chapter I am particularly interested in the extent to which the responsibility of the individual citizen-consumer holds true when approaching consumption from a practice theoretical perspective, that is, in fact, promoting the routine character of day to day activity in the procedural steps associated with different practices. Put differently: how can an understanding of social practices, as opposed to the traditional individualized understanding of the consumer as market participant, account for responsibilities in relation to a societal problem such as food waste. Looking at different systems or practices of provision and accordingly fighting food waste, the "foodsharing" initiative serves as an empirical case, being politically motivated and aiming to change systems of provision in particular and achieve social change in a broader sense. Building on that, the question must be asked what potential for change is accompanied by lifestyle politics and political consumption in terms of social change and social movements.

Baringhorst (2015) states: "lifestyle practices are still largely blind spots of social movement and political participation research." The relationships in the nexus of politics, lifestyle and social movements are still poorly understood and recognized, and not empirically researched, especially as far as collaborative consumption is concerned. One of the few exceptions is the study by Forno and Graziano (2014) showing how the collectivity in community organizations pleads for change through consumption. The point here is to show the blurring of boundaries in lifestyle-related exposures and overlaps with work. This chapter investigates the "foodsharing" initiative from the perspective of a lifestyle movement (Haenffler et al. 2012; Wahlen and Laamanen 2015). The ideas of Forno and Graziano (2014) seem helpful in this context. In order to understand sustainable community movement organizations, Forno and Graziano (2014) distinguish between, on the one hand, organisational resources and repertoires of movement action and, on the other, consumer culture and identity. Along these two lines of argumentation, the core of what follows analyses the case study of the "foodsharing" community. This chapter accordingly views contemporary collaborative consumption as undermining the traditional idea of social movements as political action opposed to the private realm of everyday life (cf. Haenffler et al. 2012). In what follows, the "foodsharing" case study will help to illuminate aspects of individual and collective action in contemporary collaborative consumption. Moreover, "foodsharing" is seen as being of particular interest in so far as the growing community and their activities have given rise to a political conflict with the food safety authorities in Berlin.

Methodological approach

This chapter is a case study of the "foodsharing" initiative in Germany. The initiative seeks to redistribute food which is still edible and make it available to consumers. Excess food is provided primarily by placing food in public distribution points such as cabinets or refrigerators, in so-called "fair-dispensers" (*Fairteiler* in German, based on the verb *"verteilen"* meaning "to distribute", hence connoting fair distribution of food that is still edible). Around 300 of these distribution points exist in Germany, of which 25 are located in Berlin. In January 2016, the local food authority in Berlin announced that the provision of food in public "fair-dispensers" does not meet public food safety requirements. This decision led to conflicts between the "foodsharing" initiative and the local food authority. This chapter examines these incidents in the general context of the politicization of "foodsharing" consumption practices in individual and collective action, as manifested in community structures and related consumer identities, in particular the positioning of "foodsharing" as non-conformist collaborative consumption and provisioning practice vis-a-vis the food safety authorities.

The case study approach here used is an open, empirical approach to the "foodsharing" initiative, not limited to obtaining responses through methodological requirements (cf. Yin 2014). The case study uses and applies various data, methods and techniques which contribute to examining the "foodsharing" initiative from different perspectives. This method scrutinizes naturally occurring data along with empirical material deriving from websites of the initiative and related discussion forums with official documents such as position papers, FAQs and organizational structure information. Individual unstructured interviews and participant observation took place in February 2016. The main question in the interview with a representative of the initiative pertained to the role of the consumer in "foodsharing", in order to understand how the community is organized. Further analysis was based on the distinction that Forno and Graziano (2014) make between community structure and consumer culture in researching sustainable community movement organizations.

Foodsharing as community movement organisation

As outlined above, food waste may be considered as a tangible societal problem which in the empirical case of "foodsharing" may help us to reflect on individual and collective action and the role of trust. We may theorize further on politicized contemporary collaborative consumption practices and distinguish empirically between the individual and the collective in community movement organizations. Theoretically I draw upon the ideas of Forno and Graziano (2014) which highlight on the one hand organisational resources and repertoires of movement action and on the other hand consumer culture and identity (see also Bennett 2012). Furthermore, a conflict between the initiative and food authorities assists in reflecting on the political nature of contemporary collaborative consumption communities. Accordingly, the question arises how activities in the sustainable community movement organisation surrounding "foodsharing" can be theoretically categorized as individual or collective action, in a critical assessment of individualized collective action.

Community structure, organisational resources and repertoires of action

Origins of the "foodsharing" initiative date to late 2011 and 2012. The principle idea underlying "foodsharing" is that there is too much food being wasted and that individuals can become active in fighting food waste. Preserving food from being disposed of can be done at different levels, as outlined above. Dumpster diving and individuals looking for edible food in waste containers of supermarkets can be considered the initialisation of "foodsharing". The start of the initiative can be considered to coincide with the first collaborations with supermarkets which agreed to have individuals pick up food, in order to prevent illegal dumpster diving. Currently, the initiative collaborates with around 3,000 companies such as supermarkets, canteens and other businesses involved in systems of provision.

Members of the "foodsharing" initiative visit these companies regularly at agreed times and places in order to pick up food that would otherwise be disposed of. After pick-up, the excess food is placed in public distribution points such as cabinets or refrigerators, in so-called "fair-dispensers" (see above), where other individuals can come and pick up the food. These distribution points are publicly accessible, and there are rules concerning what sort of food to place in them, hygiene matters, 'best before' dates, forbidden foodstuffs and other aspects. Another, less often used option of the initiative is to offer food in so-called "food baskets", through a web platform or mobile phone application. Here individuals or families can get rid of excess food, for example when they have bought too much or before leaving on holiday.

According to the initiative, almost 20,000 volunteers have saved 5 million kilograms of food waste from disposal to date. As these numbers demonstrate, there are numerous participants involved in the initiative. In order to organize the practices of the initiative, "foodsharing" has in its short life since 2012 developed a detailed and elaborate community structure. This structure should guarantee the stability of activities which aim to reach the goal of reducing or even abolishing food waste. In that way, activities and practices in the "foodsharing" initiative may be seen as part of a political lifestyle or in terms of a community of volunteers who speak of a lifestyle movement. In the movement activity, individual, private and alternative practices of food provisioning are moving to the fore.

In contrast to the traditional understanding of social movements, in which collective and public activity dominates the discourse of collectivist collective action (Stolle and Micheletti 2013), "foodsharing" comes across as mostly alternative, private activity, yet supported by commercial market activity. The collectivity of "foodsharing" becomes evident in the community structure: existing rules and regulations, for instance what food and how to place it in cabinets and refrigerators as well as ways of administering the community in order to guarantee stability, e. g. regular meetings and reflecting the organisational structure. These activities also increase the role of trust, because social relations within the community are important. Hence, considering "foodsharing" as a contemporary collaborative consumption movement, members occasionally collectively express their involvement in the initiative, but usually are not involved in explicit associations such as a political party or registered associations.

The social change advocated is essentially not manifested in the collectivity of the organization, but in the social practices of individual participants picking up and sharing food with others. Accordingly, "foodsharing" primarily aims to raise awareness and change the consumption practices of individuals and, at a secondary level, beyond the consumers, also those of companies which co-operate. It is thus not surprising that the initiative cooperates with supermarkets which have a more sustainable assortment of produce, such as organic supermarkets. The initiative as a community supports consumption practices in everyday life such as procurement, supply, processing and disposal of food, counteracting commercialized business practices which produce food waste. Here it should also be stressed that there is a variety of social practices that are relevant in the community. The focus is not on market withdrawal, but on alternative practices of provision.

The "foodsharing" initiative operates under a relatively strictly regulated yet open organizational community structure and draws on a variety of organisational resources and distinct repertoires of action. The initiative is very open to outside participants. In fact, everyone can participate: even without being registered on a website or a platform, it is possible to take food from the food distribution points. The initiative is thus organized with a very low entry threshold. Nevertheless, in terms of repertoires of action, there is a clear set of roles or positions involving characteristic responsibilities, but without promoting a strict hierarchy.

To start with, individuals can, as the initiative states, become "food sharers". In that starting position, individuals get acquainted with each other in the organisation and are able to give and take food, either in food baskets or by taking food from the public dispensers. In this way, repertoires of action extend beyond organisational resources. This ease of access is both an advantage and a disadvantage, as I will show below. Nevertheless, it is easy to become involved in preventing food waste. Without promoting a particular hierarchy, the "foodsharing" organisation has introduced different roles and organisational resources which combine with related repertoires of action to increase trust between activists. Beyond being a "food-sharer", it is possible to become a "food-saver", a person responsible for picking up food from a particular company, an ambassador, or to become involved in one of the existing working groups.

Within the role of "food-saver", participants do not just place food in the distribution points or offer food baskets, but are also entitled to participate in food rescue operations. The reality of rescuing food is explained by a participant in the "foodsharing" initiative as follows:

> We, as "foodsharing", go to companies and make a so-called cooperation agreement with them. This means that they get a disclaimer from us, they are no longer in charge of the food. And we want not only to make food available to the people participating in foodsharing, but also to those that otherwise cannot access food, or only under certain restrictions because they have no money, or are living on the street. There are still people who do not want to go to foodbanks, where you have to keep a record to show that you are in need, you have to give money at foodbanks, even if it is very little. And we are in this respect very decentralized. Our distributors are really integrated in the cityscape, in a niche, or as now, in our backyard. Very decentralized … but that is the idea,

> to save food. A lot of food is indeed thrown away in the household, and our distributors are also for the neighbourhood. So people who go on holiday or have leftovers, for example those who have bought too much, are also invited to put food there. […] It is really used by private people who are not really in "foodsharing". This is actually our idea, that we spread this idea among the population: foods do not belong in the garbage, but you should be able to pass them on, if they are not consumed by you. (foodsharing representative on 17 February 2016, own translation)

"Rescue operations" are visits to companies, as outlined above, where "food-savers" pick up excess food that would otherwise be wasted. They may take as much as they want or can eat from the food they rescue, and hand over the remaining food to the food distribution points. In order to become a "food saver", participants must fill in an online questionnaire, attend local meetings and take part in several food rescue operations with more experienced members, in order to get acquainted with the procedures, get to know the company and the locality where food is saved:

> First you need to complete the questionnaire for becoming a food-saver. Then you will be contacted by an ambassador of your region, to let us know that you are interested in picking up food from the companies. You will then receive suggested dates for the first joint rescue action which you are required to confirm. The introductory pickups should provide the possibility of getting to know foodsharing practically. […] After three successful introductory rescue sessions you get your FoodSaver passport (if the conditions are met). Then you will be added to our planning as a food-saver with the companies of your choice. (foodsharing, 2016, own translation)

More experienced "food-savers" may become responsible for visiting a particular company. As when becoming a "food-saver", there is an online questionnaire, and it is recommended to work jointly with others at first, before taking on sole responsibility. It is possible to grow even further in the organisation and become a "foodsharing-ambassador". Ambassadors are responsible for accrediting new "food-savers" and organising meetings and exchange between "food-sharers" and "food-savers". Beyond that, ambassadors are involved in finding and recruiting new companies and coordinating the activists responsible for the companies. Here again, social relations are very important to the organisation: a participant can grow as others' trust in that person grows.

Repertoires of action are not limited to "foodsharing" activities as such. There are also (regional) working groups and task forces which discuss and implement community organisation e. g. on IT support, supply chains, events, organising distribution points, communication materials (explanatory videos, administration, etc.), public relations (social media, wiki, etc). Finally, there is also a group on organisational community structure looking after how particular roles are allocated, how to achieve greater transparency, how to enhance the principle of subsidiarity, how to divide power across those involved and how to stay flexible and operationally effective.

The different repertoires of action enable reflection on the adjustment of the relationship between individual and collective action (cf. Baringhorst 2015). Overall, the general objective of the initiative can be situated primarily in environmental protection, which "characterizes a tension between individual and collective dimensions of action" (Baringhorst 2015, p. 20), as individual action is assumed to harm the (collective) environment. Hence, it should be noted with respect to "foodsharing" that individualized activity is often situated in a collective framework. The role of individuals as "food-savers" supplying food and those entitled "food-sharers" on the receiving end are not in opposition: often suppliers and users are found in the same person, who acts in their own self-interest while simultaneously considering the community. The question arises as to whether and where to draw the exact boundaries between individual and collective aspects of social practices. Discursively they are partially differentiated in roles or functions, but how this affects the activities of those involved cannot finally be answered with the available data. There are likely to be blurred borders and smooth transitions between suppliers and users, but also between the individual and the collective.

In further repertoires of action in systems of provision, "food-savers" also assist the companies involved, by sorting food that is still edible and not wasted, in the end also reducing company costs for food that they do not have to dispose of. Participants in "foodsharing" thus take responsibility in their role as "food-savers". Here too the role of trust is imperative: sorting food which is still edible, deciding what is to be placed in the public dispensers, requires trust in those doing the job. Moreover, in these practices, a moralisation and politicization of everyday life comes into play: conventional commercial provisioning practices are not questioned, but the initiative seeks to demonstrate possible alternative modes of provision. Participants in "foodsharing" do not necessarily participate because of monetary aspects influencing their quality of life, but because of non-monetary aspects outside the capitalist system combined with opportunities to satisfy basic needs. As such, individual participants in "foodsharing" are part of the private and individual lifestyle in the collective "foodsharing" community which takes on responsibility for reducing food waste.

The social practices approach comes back into play, not only with regard to the supply process, but also ontologically in connection with acceptance and recognition of alternatives similar to dumpster diving and waste-derived food in a welfare society. Repertoires of action pertain to the participation of individuals in the initiative and its inherent practices. For example, organisational resources have been developed and, as outlined above, organisational learning has been implemented to ensure, for instance, appropriate storage of food in the public food distribution points and trusting participants to act according to food safety guidelines. Together with food inspectors, usage guidelines have been developed in order to guarantee food safety. For example, no products with "use by ..." – date may be placed in the public distribution points. Furthermore, there are counsellors advising on the management of the public refrigerators, as well as giving tips on foods which are (or are not) sensitive or particularly sensitive. There are also guidelines on the transport, the expiry date and the use-by date of foodstuffs. Sensory aspects of the food and possible related diseases are also considered. The distribution points are cleaned daily using a specific method to prevent contamination of food.

Turning to the conflict of the "foodsharing" community with the public food authorities that has been mentioned above, we may reflect upon individuality and collectivity with regard to community structure and repertoires of action. The community has been growing quickly, and so has attracted attention from the media, governmental institutions and individuals. For example, the "foodsharing" initiative has won an award from the Federal German Ministry for Food and Agriculture, to highlight innovation with regard to the reduction of food waste. This award can be seen as a consumer policy measure in support of the ideas underlying "foodsharing". On the other hand, the "foodsharing" initiative is in conflict with the local food authorities in Berlin. In the latter case, the local food authority, being responsible for food safety, did not see how the associated risks, e.g. in relation to hygiene, are adequately addressed. The food authorities therefore attempted to classify the individuals, and hence the community, as a business, which implies different food safety standards in terms of responsibility, traceability and documentation. The food authorities attempted to make clear the organizational diffuseness and the repertoires of action of the "foodsharing" initiative and thereby define more clearly its blurred boundaries.

Concerning community structure, organisational resources and repertoires of action, we may conclude that the "foodsharing" initiative manifests as lifestyle politics in a contemporary collaborative consumption movement. Consumers consciously and actively adopt a lifestyle, the primary goal of which is social change to reduce food waste. Here the consumption practices approach is deemed useful because it is not primarily concerned with the consumer's market withdrawal, but rather with alternative food provisioning practices. Repertoires of action in the community may thus help to understand politically motivated activity. Further, these practices assist in reflecting on the role of individuality and collectivity: consumers do not necessarily participate to satisfy their individual needs, but also to meet the collective goal of reducing food waste.

Consumer Culture and Identity

Individualist collective action in the lifestyle politics of "foodsharing" also relates to the question of how individuals identify with the community. What kinds of identities are constructed? The organization of "foodsharing" is set up to be easily accessible in the first instance, and providing foods does not necessarily involve a registration with the community, a website or elsewhere. As becomes apparent from the community structure, organizational learning towards identification with the initiative is implemented in the way repertoires of action are spread across it. Hence, membership is likely to build on the basis of a "foodsharing" identity, negotiating individualist collective action through participation (cf. Bennett 2012).

Participants in the "foodsharing" initiative supply food for themselves and others through their activity, hence reducing food waste. In this way participants contribute to sustainable development, and hence to social change. "Foodsharing" reveals both anti-materialistic and anti-commercial stances in relation to conventional food supply in and by (super-)markets. For example, one of the founders of "foodsharing" stated his goal was to escape the capitalist system and to live his everyday life without commercial and market dependence. These aspects pertain to the identity of participants in "foodsharing" in different ways.

The impetus for consumption practices in "foodsharing" is in line with traditional social movements, such as the environmental movement and the women's movement: reducing food waste is largely ecologically motivated and aimed at ensuring equal access to food. Traditional social movements pertain to what Stolle and Micheletti (2013) denominate collectivist collective action. The community has set itself as a goal to influence social conditions in the systems of provision. In collectivist collective action, certain conditions are condemned. Political, commercial or other social actors set out to influence them and bring about change, and political opportunity structures therefore come into play. Structures of authority are challenged and questioned in order to transform a system dominated by conventional market players.

Unlike traditional social movements, lifestyle politics (de Moor 2016) and lifestyle movements (Haenffler et al. 2012; Wahlen and Laamanen 2015) focus on cultural objectives. The "foodsharing" initiative seeks change in the broader food culture, in the direction of less food waste. For example, the above-mentioned "fetishism of immaculacy", with food being thrown away at various points in the food chain, is accompanied by a consumption ethos which accepts only food that does not have any sensory impairments. This aspect of the food culture is questioned by "foodsharing": foods which are no longer considered spotless by food retailers or canteens will be picked up by the initiative and handed over for consumption by participants and made available to other members. The initiative thus aims to reduce food profligacy. Accordingly, alternative provisioning practices are supported and in some cases replace acquisition routines of everyday life by the non-commercial supply of "foodsharing".

It might be argued that monetary aspects play a role in "foodsharing", particularly with regard to the provision of food to less well-off consumers. The "foodsharing" initiative communicates that the threshold for access to food is low, because food is freely available without registration, unlike at foodbanks. Hence, "foodsharing" might be perceived as stigmatizing consumers less (cf. Van der Horst, Pascucci and Bol 2014). This is explicitly formulated by "foodsharing" as follows:

> We would avoid declaring participants of foodsharing as needy, because that is associated with the risk of stigmatization. Foodsharing is possible for everyone, regardless of origin, social status or belief, in order to have access to rescued food. […] Certainly needy people can cover part of their demand for food through foodsharing, but there is no security of supply. A long-term goal of foodsharing is to reduce food waste. Therefore, it is important to us that someone can live by sharing food, but no one goes from an emergency situation to dependency on foodsharing. (foodsharing 2016, own translation)

The "foodsharing" initiative identifies participants not as consumers but as "food-savers", "food-sharers" and "foodsharing-ambassadors". It focuses on the individual level, where individual participants want to change procurement practices and make a difference in order to achieve social change. Nevertheless, although the objective is focussed on individuals, the community as a whole collectively aims at the reduction of food waste on a societal level and advocates changes in the food culture which are closely intertwined with personal identification of the initiative as an alternative way of providing food.

The current food culture should, according to "foodsharing", be changed in order to conserve resources. As part of the everyday food culture, food is consciously and unconsciously wasted in households and families (Aschemann-Witzel et al. 2015). From a practice theoretical perspective, emphasizing the routine character of conduct, the social meaning and common understanding of food are relevant in terms of identity. Identifying with the "foodsharing" initiative and the pro-environmental stance is certainly germane to a green identity. Saving food then relates to saving the world. Hence, eating food that is still edible that otherwise would have been thrown away is making a change in the individual context, but also extends to the overall allocation of food. Take for example the 'best before' date: it might be seen as a policy instrument to safeguard consumers. It seems, however, that the date is exploited by food producers to increase sales, in the sense that food beyond that date is often edible and harmless to health:

> There is an expiration date for Himalayan salt, which is stored millions of years in the mountains, but when it is packaged, it must get an expiration date for the consumer. (Personal conversation with foodsharing representative on 17 February 2016)

"Foodsharing" thus questions the expiry date and tries from its perspective to contribute to the sense and nonsense of government regulations and the associated practices initiated by the food manufacturers. In this sense, "foodsharing" is political, aiming at the individual practices of participants in order to change the food culture, hence also food manufacturing practices and the associated regulatory framework.

Turning to the conflict with the local food authorities in Berlin, it can be said that the authorities follow the objectives of implementing legal requirements and policy objectives for the protection of consumers. Here the transition between cultural and political objective becomes evident. Although the primary objective of the "foodsharing" initiative must be regarded as cultural, in the sense of change in the food culture, the political dimension becomes clear when the authorities challenge "foodsharing" practices. The locus of politicization of "foodsharing" can be found in the suitability of the legislation (or its implementation), or to be more precise, in the changing societal framework, where food waste is becoming a bigger problem than food safety. Furthermore, the political nature of this case is evident in the associated power structures. As becomes clear from the related media debate and the personal interview on the specific case, the food authority has used its power to nearly shut down the initiative, which from a participatory perspective should rather lead to an optimization of requirements (see de Moor 2016). Accordingly, the question whether the political in collaborative consumption strengthens or weakens democracy (Lamla 2015) will not be answered here, but from a participatory perspective the problem should be solved in a deliberative exchange.

The "foodsharing" initiative corresponds to the idea of a collaborative consumption movement insofar as rescuing food takes place with a certain regularity, building on the identity of participants. In the personal interview it became clear that, for example, in the context of Berlin, participants in "foodsharing" visit supermarkets on a daily basis. Participants are prepared for continuous activity in the "foodsharing" initiative. The aim of the "food-saver" training process is to sensitize participants to handling the food on a very practical level, and to clarify and accept responsibilities inherent in the activity and to develop trust in the community. Responsibility is relevant for two reasons. On the one hand, "food-savers" are responsible for the quality of food placed in the distribution points. On the other hand, they need to be aware of the social responsibility which goes with the (re-)distribution of food through the "foodsharing" community.

Concluding remarks

The field of social movement research has opened up to politicized lifestyles and politically motivated consumption practices (Baringhorst 2015; de Moor 2016; Dubuisson-Quellier 2015; Haenffler et al. 2012; Wahlen and Laamanen 2015). On the other hand, contemporary forms of collaborative consumption are being considered an arena where individual consumers as well as (virtual) consumer communities are working towards social change (Wahlen and Laamanen 2017). The "foodsharing" initiative has been discussed in the case study contained in this chapter to reflect on individualist collective action and the blurring of boundaries between social movements in the traditional sense and lifestyle movements on the other. We have also considered the role of trust in this initiative. Whereas collaborative consumption movements are largely understood as private acts, we should emphasize that "foodsharing" is at times also a public activity, as the episodic participation in the case of the public confrontation with the food authorities showed. In this sense, the individualized everyday world of consumers in "foodsharing" is also addressing the social problems of food waste. The food culture is the cultural objective which makes "foodsharing" a politically motivated activity. Politicization of the consumer sphere may therefore contribute to increased democracy (Lamla 2015), but only if participation achieves results in the form of decision-making (de Moor 2016).

For consumer policy, it can therefore be noted that initiatives of a political collaborative consumption movement organisation draw on a variety of repertoires of action in their community structure, which need to be acknowledged in a participatory process in policy decision-making in order to realize the full potential for change. Accordingly, a discursive process might assist in counteracting obstacles to the wider acceptance of social change. To change social practices and food cultures, there is still a need to integrate implicit knowledge on the impeccability of foodstuffs. Sensory skills for testing the food play a role in "foodsharing", as does the knowledge of alternative practices in everyday life. Nevertheless, associated identities in the process of social change also seem to play a significant role, along with trust in the community. For those who see food as supposedly non-political, "foodsharing" as an initiative and case study is also non-political, and may be located between the private and the public spheres. Polarization in connection with food waste makes clear how individuals and communities act in these spaces and in this way show political commitment. However, it is questionable, as Lamla (2015, p. 34) aptly argues, whether the "rudiments of alternative practices, public experimentation and collective learning currently have the power to produce new arrangements in the hierarchical relations of institutional securities and of justification". Consumer policy can and should contribute to and support initiatives aimed at moving social change in a more sustainable direction, as this case study has demonstrated.

References

Aschemann-Witzel, J., de Hooge, I., Amani, P., Bech-Larsen, T., & Oostindjer, M. (2015). Consumer-Related Food Waste: Causes and Potential for Action. *Sustainability,* 7: 6457–6477. doi:10.3390/su7066457.

Baringhorst, S. (2015). Konsum und Lebensstile als politische Praxis – Systematisierende und historisch kontextualisierende Annäherungen. *Forschungsjournal Soziale Bewegungen – Analysen zu Demokratie und Zivilgesellschaft,* Heft 2 (Juli), S. 17–27.

Bennett, L. (2012). The Personalization of Politics. Political Identity, Social Media, and Changing Patterns of Participation. *The ANNALS of the American Academy of Political and Social Science,* 644 (1): 20–39. doi:10.1177/0002716212451428.

De Moor, J. (2016). Lifestyle politics and the concept of political participation. *Acta Politica,* advance online publication, 8 January 2016. doi:10.1057/ap.2015.27.

Dubuisson-Quellier, S. (2015). From targets to recruits: the status of consumers within political consumption movement. *International Journal of Consumer Studies,* 39, Nr. 5: 404–412. doi:10.1111/ijcs.12200.

FAO (2013). Food waste and food loss. http://www.fao.org/food-loss-and-food-waste/en/ (accessed: 12.8.2016).

foodsharing (2016) Internet source: https://foodsharing.de/ (accessed 12.8.2016).

Fridrich, C., Hübner, R., Jaquemoth, M., Kollmann, K., Piorkowsky, M.-B., Schneider, N., Tröger, N., & Wahlen, S. (2014). Bamberger Manifest für ein neues Verbraucherverständnis. *Journal für Verbraucherschutz und Lebensmittelsicherheit,* 9 (3): 321–326. doi:10.1007/s00003-014-0880-1

Fridrich, C., Hübner, R., Kollmann, K., Piorkowsky, M.-B., & Tröger, N. (2016). *Abschied vom eindimensionalen Verbraucher.* Wiesbaden: Springer VS.

Haenfler, R., Johnson, B., & Jones, E. (2012). Lifestyle Movements: Exploring the Intersection of Lifestyle and Social Movements. *Social Movement Studies,* 11, 1: 1–20. doi:10.1080/14742837.2012.640535.

Kjaernes, U., Harvey, M., & Warde, A. (2007). *Trust in Food: A Comparative and Institutional Analysis.* Basingstoke: Palgrave Macmillan.

Lamla, J. (2015). Politischer Konsum und (Post)Demokratie – ein Debattenbeitrag. *Forschungsjournal Soziale Bewegungen – Analysen zu Demokratie und Zivilgesellschaft,* Heft 2 (Juli), S. 28–36.

Moor, J. de (2016). Lifestyle politics and the concept of political participation. *Acta Politica,* online first. doi:10.1057/ap.2015.27.

Niehof, A., & Wahlen, S. (2017). Moralities of sharing and caring – Gender and food in the moral household economy. In B. Bock & J. Duncan (eds.), *Gendered food practices from seed to waste.* Yearbook of Women's history, 36: 147–161.

Quested, T., Marsh, E., & Stunell, P. (2012). Spagetti soup: The complex world of food waste behaviours. *Resources, Conservation and Recycling,* 79: 43–51. doi:10.1016/j.resconrec.2013.04.011.

Stolle, D., & Micheletti, M. (2013). *Political Consumerism. Global Responsibility in Action.* Cambridge University Press

United Nations (2016). Sustainable development goals. Internet source: http://www.un.org/sustainabledevelopment/sustainable-development-goals/ Accessed 8 February 2017.

van der Horst, H., Pascucci, S., & Bol, W. (2014). The "dark side" of food banks? Exploring emotional responses of food bank receivers in the Netherlands, *British Food Journal,* Jg. 116, Nr. 9: 1506–1520. doi:10.1108/BFJ-02-2014-0081.

Wahlen, S., & Laamanen, M. (2015). Consumption, Lifestyle and Social Movements. *International Journal of Consumer Studies,* Jg. 39, Nr. 5, S. 397–403. doi:10.1111/ijcs.12237.

Wahlen, S., & Laamanen, M. (2017). Collaborative Consumption and Sharing Economies. In M. Keller, B. Halkier, T.-A. Wilska & M. Truninger (eds.), *Routledge Handbook on Consumption.* Oxford, UK: Routledge.

Wahlen, S., & Winkel, T. (2017). *Household food waste.* Elsevier Compedium for Food Sciences. doi:10.1016/B978-0-08-100596-5.03368-0

Warde, A. (2016). *The Practice of Eating.* Polity Press, Cambridge, UK.

Winkel, T.; Wahlen, S., & Jensen, T. (2016). Food waste: orchestrating practices of eating. Paper presented at the Nordic Conference of Consumer Research at Aarhus University.

Yates, L. (2015). Rethinking Prefiguration: Alternatives, Micropolitics and Goals in Social Movements, *Social Movement Studies: Journal of Social, Cultural and Political Protest,* 14, Nr. 1: 1–21, doi:10.1080/14742837.2013.870883.

Yates, L. (2016). Sharing, households and sustainable consumption. *Journal of Consumer Culture,* online first. doi: 10.1177/1469540516668229.

Yin, R. (2014). *Case Study Research – Design and Methods.* Sage Publications: Thousand Oaks, USA. 5. Edition.

Riding free-riders? A study of the phenomenon of BlaBlaCar in Italy

Francesca Setiffi and Gian Paolo Lazzer

Abstract

The chapter analyses the case study of BlaBlaCar, which is a for-profit online platform offering carpooling services to connect people who need to travel with drivers who have empty seats. The carpooling offered by BlaBlaCar partially constitutes an act of contemporary collaborative consumption (CCC) because despite the brokerage company involved, which should lead us to count this practice of consumption among typical sharing economy activities, some of those involved in the ride sharing feel they are part of a community that advocates an alternative way of life to the classic model of private car use. Our empirical research has been carried out to understand the meanings and practices within the 'social world' of BlaBlaCar users. We investigate the evolution of carpooling to understand whether and how users' carpooling experiences fluctuate over time. To reach these goals, we conducted 70 semi-structured interviews with people between 18 and 35 years old who had interacted with the collaborative platform at least twice. All the users involved in the fieldwork lived in the Veneto region of Italy, and the data were collected in 2015. We used the ATLAS.ti software to elaborate on the data for content analysis. The most important result of the research is the clarification of the relationship that drivers establish with the figure of the 'stranger'.

Zusammenfassung

Das Kapitel analysiert in einer Fallstudie die gewinnorientierte Online-Plattform BlaBlaCar. Die Plattform organisiert Fahrgemeinschaften: Autofahrer mit leeren Sitzplätzen haben die Möglichkeit diese über eine Online-Plattform zur Verfügung zu stellen. Die von BlaBlaCar angebotenen Fahrgemeinschaften stellen teilweise einen Akt des kollaborativen Konsums dar. Trotz des beteiligten Maklerunternehmens der Online-Plattform fühlen sich Teilnehmer der Mitfahrgelegenheit als Teil einer Gemeinschaft,

I. Cruz et al. (Hrsg.), *Contemporary Collaborative Consumption*,
Kritische Verbraucherforschung, https://doi.org/10.1007/978-3-658-21346-6_5

die eine alternative Lebensweise gegenüber dem klassischen Modell der privaten Pkw-Nutzung bevorzugt. Unsere empirische Forschung wurde durchgeführt, um Bedeutungen und Praktiken innerhalb der „sozialen Welt" von BlaBlaCar-Nutzern zu verstehen. Die Entwicklung von Fahrgemeinschaften wurde untersucht, um zu verstehen, ob und wie sich Fahrgemeinschaften im Laufe der Zeit verändern. Empirisch wurden 70 semistrukturierte Interviews mit Personen zwischen 18 und 35 Jahren durchgeführt, welche mindestens zweimal mit der Online-Plattform interagiert hatten. Alle an der Feldarbeit beteiligten Benutzer wohnten in der Region Veneto in Italien, und die Daten wurden 2015 erhoben. Mit Hilfe der Software ATLAS.ti wurden die Daten durch eine Inhaltsanalyse analysiert. Das wichtigste Ergebnis dieses Kapitels ist die Klärung der Beziehung, die der Fahrer mit der Figur des „Fremden" eingeht.

Keywords

carpooling, collaborative consumption, Italy

Introduction[1]

The sharing economy, collaborative consumption (Botsman and Rodgers 2010; Wahlen and Laamanen 2017), connected consumption (Schor 2014; Shor and Fitzmaurice 2015) and access-based consumption (Bardhi and Eckhardt 2012) are different ways of describing the relevant change concerning the relationships between production, distribution and consumption. The definitions embrace different but contiguous dimensions of the phenomenon: collaborative consumption concerns product service systems, redistribution markets and the collaborative lifestyle (Botsman and Rodgers 2010); connected consumption emphasises the social and digital aspects of the phenomenon and is 'predicated' on peer-to-peer relationships (Schor and Fitzmuarice 2015), while access-based consumption transactions may be market mediated but avoid the transfer of ownership. 'Product' consumption refers to the time not the item.

In highlighting the different aspects of the collaborative economy, there is cognisance of the central role played by technology in making possible the sharing of goods, services and knowledge outside the lines of kinship and friendship circuits.

1 While this chapter is the result of several discussions between the authors, Francesca Setiffi wrote sections Introduction, Transportation and consumption, The BlaBlaCar case study, Methodology, The carpooling experience: savings and challenge – the consumer perspective and Gian Paolo Lazzer wrote the remaining parts of the chapter. The reviewers provided us with invaluable help with their comments and remarks.

The spread of new technologies such as social networks and apps has, above all, profoundly changed the way we produce, distribute and consume goods and services (de Kerckhove 1997; Belk 2014; Degli Esposti 2015). While technology seems to be the pivotal element characterising the majority of the new forms of sharing, the sharing economy (SE) includes many consumption practices and cannot be considered as a homogeneous phenomenon. Indeed, sharing practices have different aims, values and forms of organisation. Contemporary collaborative consumption (CCC) can be included as one of the most important expressions of the SE. It seems to be, above all, that which focuses on specific values that deeply influence sharing practices (Wahlen and Laamanen 2017).

In fact, these values are related to forms of consumption such as consumer activism and alternative lifestyles. Among others, attention to sustainability is a clear example of this value. While consumer-oriented behaviour has expanded to include the environment in consumers' purchasing choices, the culture of sharing of goods, spaces and services influences the strategies of national government and businesses, thus acting on the habits of citizens and consumers in promoting responsible behaviour from an environmental perspective (Cruz and Katz-Gerro 2016).

This research presents a case study – BlaBlaCar – which describes the beginning of the possible trajectories of CCC. BlaBlaCar is an online platform offering carpooling services, connecting people who need to travel with drivers who have empty seats. The idea of BlaBlaCar is very similar to hitch-hiking, but the Web has brought it to the market. It is a for-profit firm and is based on safe-sharing practices. It also guarantees a high level of trust among users, even though they do not know each other. The carpooling offered by BlaBlaCar partially constitutes an act of CCC because despite the brokerage company, which should lead us to count this practice of consumption among typical SE activities, some of those involved in the sharing with 'hitchhikers' feel that they are part of a community that shares an alternative way of life to the classic model of private car use. The line between economic and extra-economic factors in the case of BlaBlaCar appears to be particularly blurred. The empirical research carried out to understand the meanings and practices of the 'social world' of BlaBlaCar users is based on a theoretical framework inspired by the reflections of Wahlen and Laamanen (2017) and Schor (2014) on collaborative and connected consumption, while the so-called theories of practice (Warde 2014; Shove et al. 2009) have helped us investigate the evolution of carpooling and to understand whether and how users' carpooling experiences fluctuate over time.

Consumers' primary motivation for using BlaBlaCar is cost savings, and even if they declare that environmental protection is a relevant value, they do not choose carpooling to reduce their environmental footprint. However, the challenge of meeting strangers and believing in their honesty is not an obvious research finding. These findings are closely related to the concept of trust. In fact, the study has proven that trust – the fundamental component in the establishment of the BlaBlaCar network – changes over time and that it depends on users' experiences. It is not just a question of service; users have underlined alternative reasons for BlaBlaCar's utility. After an initial development of mutual trust between strangers, a community of users has been established. The community spirit is the fundamental base for the development of different practices. In particular, it is fundamental for the growth of CCC that it is not influenced by economic reasons alone, as is the case in purely monetary economies (Simmel 1900). In broad terms, CCC has the potential for deep social impact. The SE connects consumers who often meet in the market as strangers, but in some cases, as we will see in Section Section "Understanding the BlaBlaCar world through the lens of contemporary collaborative consumption. A community of strangers" , they can become a 'kind of community' in which relationships between members are not just economic in nature.

Along with the introduction, this chapter is divided into eight sections. The second section *(transportation and consumption)* outlines the relationship between transportation and consumption; the third section *(the BlaBlaCar case study)* deals with the description of the case study; the fourth section *(methodology)* explains the methodology; the fifth and sixth sections *(the sharing economy perspective: make strangers believe in one another – the enterprise perspective and the carpooling experience: savings and challenge – the consumer perspective)* explore the role played by the online platform and the meanings and motivations connected to carpooling; the seventh section *(Understanding the BlaBlaCar world through the lens of contemporary collaborative consumption – a community of strangers)* analyses the carpooling experience through the lens of CCC, while the conclusion discusses further lines of inquiry.

Transportation and consumption

What is different about today's SE is that the sharing space – due to increased acceleration via new media – is open not only to networks of friends or relatives but also to strangers. Analogous to what is happening in Italy with the second-hand market, which is overcoming social stigma and developing an autonomous meaning (Secondulfo 2016; Setiffi and Lazzer 2016), car sharing is also increasingly associated with a specific meaning when compared to the individual use of a car (Cruz and Katz-Gerro 2016). In fact, several studies explore transportation and note the shift from a macro to a micro-economic dimension. These studies underline the influence of the SE on traffic (Furuhata et al. 2013; OECD and International Transport Forum 2015), the business model that characterises the sharing platform (Fellows and Pitfield 2000; Cohen and Kietzmann 2014) and the impact of the 'short-term rental of durable goods' strategies on the consumer (Fraiberger and Sundararajan 2015).

Sociological and economic studies are based on different presuppositions and reach different results in terms of analyses of the degree of development and impact on social and economic systems, yet they both accept the idea that the private consumption of a car has undergone cultural change. The analysis should thus not be limited to the idea that ridesharing is a privileged point of observation for the famous 'age of access' that Jeremy Rifkin (2000) describes, which is characterised by invisible networks that amplify social and cultural inequalities. Rather, it should be seen as a way of analysing the shift from private to collective use based on a 'community of strangers'.

Towards the end of the 1980s, Teal (1987) defined carpooling as 'the sharing of the travel to work among workers'. This definition of carpooling – even if limited to a work environment – allows us to analyse the transformation of sharing rather than its innovative consequences. In recent years, mainly thanks to new technologies, words like carpooling and ridesharing have been used to describe 'an activity of sharing among individuals that have the common need to reach a destination, not necessarily for work reasons' (Bertolin, Beria and Filippini 2016, p. 3). In the definitions that Chan and Shaheen (2011, p. 94) proposed, the sharing of a car can be:

- 'Acquaintance-based' carpooling, which is formed among families/friends or co-workers.
- 'Organization-based' carpooling requires that participants join the service with a formal membership or by simply visiting the organisation's website.
- 'Ad hoc' ridesharing, which is a form of carpooling that requires minimal relationships between participants, does not include membership and is self-organised.

BlaBlaCar is an example of the second category. Departing from the 'acquaintance-based' and 'ad hoc' ridesharing categories, the 'organisation-based' category allows us to underline the transformation of consumer motivations over time as well as their level of trust in the service and in the 'community of practices'. While the first category of analysis is based on ridesharing among people who know each other, and the third category mainly deals with self-organised ridesharing, in our case, ridesharing is based on an online platform.

The BlaBlaCar case study

BlaBlaCar was established in France in 2006 and offers carpooling services. As stated on its website: 'it is the world's leading carpooling community which enables low-cost city-to-city travel by connecting people with drivers going the same way' (BlaBlaCar, February 2017). It has radically transformed the previous concept of carpooling (in France), which began in 1958 under the name organized hitch-hiking (Vincent 2008). BlaBlaCar has 30 million users in 22 countries, and within a period of three months, 10 million individuals used the service through its online platform (BlaBlaCar, February 2017).

Few studies on carpooling and ridesharing exist in the context of Italy and France (Bertolin et al., 2016), with many focussing more generally on sharing platforms (Pais and Mainieri 2015). Most existing studies have analysed the company's business model (Di Minin et al. 2016) and user profiles (Bertolin et al. 2016; Farajallah, Hammond and Pénard 2016; Mundler et al. 2016). The user profile investigations (drivers and passengers) do not coincide in terms of descriptions of the typical consumer. They are also not comparable because they are based on samples selected on the basis of different designs. For example, Bertolin et al. (2016) analysed publicly accessible options that are only available in some Italian municipalities. Mundler et al. (2016) completed an online survey exploring the socio-demographic profiles of users, and Farajallah et al. (2016) demonstrated that experienced users can negotiate lower prices and that users with Arabic-sounding names are less likely to be chosen as drivers. Generally speaking, the average driver is a male between 26 and 35 years old, who only occasionally uses the service (Bertolin et al. 2016; Mundler et al., 2016). A research team from Scuola Superiore di Sant'Anna (Di Minin et al. 2014), which has collaborated with BlaBlaCar since 2014 by conducting periodic surveys to study user profiles (5,000 cases), reached the same conclusions. According to these findings, consumers average 31 years of age; 71 % of them are employed; and more than 60 % are based in northern Italy.

In broad terms, we can say that according to Schor (2014), four kinds of activities characterise a sharing economy: 1) the recirculation of goods; 2) an increased utilisation of durable assets; 3) the exchange of services and 4) the sharing of productive assets. Carpooling falls under the second activity. In addition to cars, homes are also part of this sphere, as they are potential goods that can be shared. Like BlaBlaCar, home swapping has been booming (Forno and Garibaldi 2015). This activity also requires basic trust in order to connect consumers.

The intensive use of a private good characterises both examples (cars and homes are used by people who are not the owners). This practice does not result in a profit, but it does facilitate savings-based economic advantages. Following Schor

Figure 1 Sharing economy – type of provider and platform orientation

		Type of Provider	
		Peer to Peer	Business to Peer
Platform Orientation	Non-Profit	Food Swaps, Time Banks	Makerspaces
	For-Profit	Relay Rides, Airbnb	Zipcar

Source: Schor (2014, p. 4).

(2014), we can categorise BlaBlaCar among such sharing activities because it utilises a peer-to-peer design and a for-profit infrastructure (see Figure 1).

Juliet Schor's scheme focuses on the platforms and analyses the aims and business strategies of category providers. Conversely, this research takes into consideration users' points of view. In doing so, it sheds some light on the evolution of carpooling-linked practices. Collaborative transportation is made up of practices that evolve faster than the providers' design: our aim is thus to consider user agency as a key factor in value-centred practices.

As we will show in the data analysis, BlaBlaCar presents a kind of 'challenge' for consumers. This involves the sharing of a consumer space as well as a travel experience, which allows BlaBlaCar to get back into contemporary forms of collaborative consumption (CCC) as 'alternative lifestyles' guided by reasons not exclusively related to the economic sphere. It is the answer to an economic need, yet it may also become a means of a kind of community building that goes beyond the economic, whereby that which is shared is not only a good or service.

Methodology

Our research is qualitative in that it aims to interpret the evolution of meanings and practices related to carpooling. The sample includes individuals who have used BlaBlaCar for work or leisure (both drivers and passengers). All the respondents have used BlaBlaCar at least twice. By using a qualitative approach, we describe the evolution of their opinions, knowledge and feelings about BlaBlaCar and about sharing a ride. We asked them to tell us about their first and most recent experiences and whether their encounters in connection with these experiences were positive or negative. We considered the use of carpooling following the practice-trajectory logic (Warde 2014). On the one hand, interviews set out to investigate consumer motivations, the storytelling of the consumption experience and the kind of social relationships that arise among strangers when travelling. On the other hand, we investigated how (and whether) the practice changed over time (from its first occurrence to its modern use). Moreover, the interviews hold that this form of consumption is intrinsically social and digital: a 'connected consumption' (Schor 2014).

In short, by adopting a qualitative approach, our research aims to investigate practices and meanings associated with carpooling. To reach this goal, we conducted 70 semi-structured interviews with people between 18 and 35 years old (both male and female) who had interacted with the collaborative platform at least twice. All users involved in the fieldwork lived in the Veneto region (Italy), and the data were collected in 2015 (October and November). We used the ATLAS.ti software to elaborate on the data for the content analysis.

The sharing economy perspective: make strangers believe in one another – the enterprise perspective

The biggest challenge for enterprises operating inside the SE has been to create a coordinated and harmonious environment within which different consumption practices can develop effectively and without too many strains (Shove et al. 2009). The SE does not always involve experts or professionals who are used to managing economic activities. This could create problems in managing supply and demand. As for the supply side, in fact, the SE involves actors who are often just amateurs – people looking for extra income or people trying to optimise a cost (as in the case of BlaBlaCar's drivers; they do not own special driving licenses or specific permissions). Nevertheless, social practices relating to the SE have been able to ensure sufficient quality, comparable to that of traditional channels. This includes the quality of goods and services, punctuality and safety (disputes related to Uber and Airbnb, for example, focus on 'rights'-related problems rather than the quality and safety of services). Furthermore, in some cases, the world of sharing is not a real alternative but rather a set of practices with a specific utility which is not comparable. This is not, however, the most important factor in the development of practices relating to the SE. Rather, the SE requires that people engage in an exchange circuit (Zelizer 2010) enabling actors to limit their perceptions of risk; that is, the exchange activities must be safe. Actors in this economy must therefore guarantee the quality of their goods and services or – in the case of SE consumers – guarantee their payment.

Open the door to a stranger, buy second-hand goods, eat something offered or share objects. These are actions that require a higher degree of confidence than normal market exchanges. Without proper security, some practices relating to the SE would never have become as widely popular as they are now. The issue of trust, therefore, is at the heart of the development of the SE, as it is not only tied to the organisations that promote it, but also depends on the practitioners themselves. In fact, there is a double mandate: one to the organisation and another to the single person with whom you interact – an unknown person who is often not a professional. To explain how it is possible to reach sufficient levels of trust, one must consider how the figure of the stranger acquires new meanings and, simultaneously, loses its negative connotations for the user. Zelizer (2007) has shown how markets (and companies) must respond to economic and rational demands while also generating more intimate and cultural offerings. We know that trusting one another is inherent in these exchanges when they take place within a small and cohesive community or within kinship networks. However, how is trust created, and how does it become stable, if the actors involved are not professionals – especially when they may not even know each other?

First, to limit risks (i. e. inherent problems in meeting a person whom you do not know), there must be a formal assurance system – legal and technological – to process the first mandate. This system is charged with providing sufficient and clear information on the person with whom you will come into contact. Thus, the service encourages the second mandate.

Simmel (1900) pointed out that only developed monetary economies transform strangers into ideal partners for an exchange; the deal is immune from strong feelings like love or hate and is guaranteed by a commonly recognised third party. Though the mediation of money guarantees the neutrality of the exchange, it is not necessary (if it is harmful) to share 'too much information' with others. Also, the SE requires a guarantor who, at least in the initial phase, fulfils the function of a proper medium. Some sharing practices (such as BlaBlaCar) are guaranteed by an organisation that creates a network of exchanges that are considered safe and trustworthy.

Many forms of SE have succeeded thanks to the implementation of platforms that certify the shared goods and services. In the case of BlaBlaCar, the company developed a system based on user feedback, descriptions, images and seniority, which are outlined on the website. Such platforms are an information infrastructure (Mongili and Pellegrino 2014) that permits and facilitates cooperation between groups and people. Technology is fundamental in this regard, but unlike expert systems (Giddens 1991) which focus on professionalism, platforms like BlaBlaCar's seem more like symbolic emblems – such as money. Trust is thus delegated to a socio-technical element (Callon 1998) which exploits the web's potential for facilitating connections. For this reason, the terms connected consumption (Schor 2014; Schor and Fitzmaurice 2015) and access-based consumption (Bardhi and Eckhardt 2012) seem to be more accurate than others in defining two key issues: technology and connections. The realisation of a trustworthy socio-technical entity is therefore a necessary condition, but it is not the only one. Defining the relationship between trust and a stranger is more complex.

Secondly, the risk should also be converted on a symbolic level. If the enterprise's aims and values are clear, as Schor (2014) has shown, then people's will and feelings are not so clear and stable. The SE enhances exchanges not only as economic opportunities but also as opportunities for cultural enrichment. The stranger becomes a value in and of him- or herself because he or she is the bearer of novelties who makes it possible to transcend borders and personal limits both metaphorically and concretely. We thus move from generalised trust to trust that addresses BlaBlaCar users in particular. While the former focuses on limiting the risk, the latter underlines that the opportunities associated with a sharing economy are not merely economic.

The capacity to generate confidence in a certain type of sharing economy cannot be explained ex post, considering its success and the number of people who use it. Analysis cannot focus solely on the enterprise nor on the technological infrastructure that supports it. As we will demonstrate in sections 6 and 7, trust is a sentiment that is not static; it evolves and changes. Trust is always based on different elements, including economic, environmental and technological as well as relational factors. These do, however, take on different roles from time to time. As for the actors, we can say that they not only share a good or service, but also a feeling of trust made up of implicit and explicit knowledge, rhetorical insight and common goals which define different kinds of practices. The sharing ethos has different meanings and aims that strongly depend on user experience.

The carpooling experience: Savings and challenge – The consumer perspective

Generally, there are three motivations for a consumer's choices in a sharing practice: savings, pollution reduction and the development of social networks (Schor and Fitzmaurice 2015). An ideological stance, which occurs mainly in the form of a peer-to-peer non-profit platform, closely follows these factors. This is also described in time bank research (Carfagna et al. 2011; Laamanen, Wahlen and Campana 2015). From the interview analyses, we can state that savings is one of the main motivations for using BlaBlaCar among those who adopt proactive behaviours in periods of crisis (Paltrinieri 2012; Secondulfo 2012). Indeed, the sample considered this form of consumption a personal challenge that equates sharing a car for travel purposes with an act of generalized trust in the whole population of users. BlaBlaCar is a carpooling service representing a market-mediated system of exchange that generates economies in one's lifestyle and allows people to challenge themselves beyond the market.

Using a car as a shared resource is a personal challenge (i. e. contacting people whom you do not know before the ride) requiring individuals to trust strangers. BlaBlaCar's site succeeds in transforming an event which may be stressful (the planning of a low-cost trip) into a lucrative activity, through a sort of 'gamification' of the car ride (Maturo and Setiffi 2016). Saving money, for example, is one of the main motives for choosing this kind of service, and it is consistent with a general consumption attitude directed to minimizing the effects of the economic crisis on personal behaviour. Moreover, interviewees – especially when narrating their first experiences with BlaBlaCar – described encounters with strangers as a way of feeling autonomous and of challenging themselves. In some cases, interviewees said they were not ready to use BlaBlaCar alone. They therefore joined BlaBlaCar with relatives and/or friends. In such cases, only the drivers or other passengers remained outside the social network. Serendipitously, a carpooling experience leads to the rediscovery of social relationships.

> It has been a way of being independent and therefore not always depending on parents for rides (…) obviously, at the beginning, we were shy because you must trust and grant trust to someone you do not know. But overall, it went fine. (P 17: Woman 18 years old)

> The first time I used it, I was with my roommate. We met a guy who was also attending university here, and by chance, we discovered that he was taking classes along with another friend. So we then became friends, and we also became friends with a woman who goes back and forth. (P 9: Man 20 years old)

> When there is a lack of money, we rediscover things that were lost. Therefore, we talk, we share and we also use social networks to go back to the ancient values of sharing and trust in others. (P 2: Woman 21 years old)

The main purpose of carpooling is to save money – even if the platform mainly focuses on the slogan, 'Save the planet'. This environmental angle, however, does not explain why people choose BlaBlaCar for the first time. Indeed, the interview analyses reveal a generally positive attitude to the environment that does not coincide with the main motivation for choosing BlaBlaCar.

In a recent quantitative study exploring these sharing practices Hamari, Sjöklint and Ukkonen (2016, p. 2054), for instance, underline how 'perceived sustainability significantly predicted attitudes of collaborative consumption while it did not have a direct association with behavioral intentions'. In the same light, another recent study has shown how household consumption sharing practices can positively affect the environment (Yates 2016). In this case, the decreased impact of environmental motivation on consumption decisions positively affects the lived experience of BlaBlaCar users, but it also plays a marginal role in the choice to carpool. The majority of interviewees underlined the economic factor as the only driver of their choice, not mentioning any environment-related reasons.

> The sharing economy allows us to share experiences and also allows everybody to save money and to preserve the environment – at least in the transportation sector (…), but savings come first of all. (P 8: Man 30 years old)

> The first reason is because it allows me to save money, and the second motivation … I do not know, but it is not because I am environmentally friendly. (P 27: Woman 20 years old)

The practices of BlaBlaCar users change over time. This can even happen within the 'walls' of the car, which the enterprise cannot plan for. Accordingly, BlaBlaCar has developed a brand that enables consumers to live the experience of sharing a car ride. Other brands do communicate a set of values that the consumer can more or less recognize – but as for products (and in the case of BlaBlaCar), the meanings that individual consumers associate with them may be different from those attached to the brand. As stated by Pais and Provasi (2015), 'blabla' refers to a kind of collaboration that strongly distinguishes the reciprocity and trust built through carpooling, configuring it as a form of socialization from the market reality promoted by the car sharing system (e. g. Zipcar).

Understanding the BlaBlaCar world through the lens of contemporary collaborative consumption – a community of strangers

Practices related to the SE and CCC are somewhat heterogeneous. In addition to the goods and services that are shared, they differ according to range of action, number of users and popularity. As already highlighted, further differences depend on whether they are based on a peer-to-peer or business-to-peer scheme and whether they are for-profit or non-profit according to the platform's aims. These differences, in the long run, show that not all sharing experiences have created real communities, because users' agency and practice are still fundamental. Some of them just formalize existing economic exchanges, despite the fact that these practices are often seen as a stimulus to community development.

Two of the most important variables for the development of a community based on these sharing practices are geographical boundaries and the frequency of exchanges between users. For example, it has been suggested that community functions are stronger if exchange practices are tied to local and spatially defined contexts (Albisson and Perea 2012). Moreover, with food sharing at a local level, it is possible to take advantage of pre-existing relationships between social actors. It is thus possible to reach high levels of trust rapidly and sustainably. Conversely, as highlighted by Schor (2014), other forms of sharing are not transformed into real communities.

In particular, some of the values conveyed by the sharing economy, such as altruism, would conflict with typical market dynamics, restricting the 'community spirit' and the possibility of moving on to a form of CCC. The data collected lead us to claim that BlaBlaCar inhabits a liminal area between a real community and the total absence of a sense of membership shared between users. Although BlaBlaCar is not intended as a territorial network defined by clear boundaries, it activates relationships that are both local and repeated over time. Above all, however, it activates processes that change the relationship with the so-called 'stranger'.

Using the theories of practice approach, BlaBlaCar can be characterized precisely by four elements that are typical in a social practice: formal rules produced by the organization; shared knowledge; users who share the same feelings and emotions; and the need for specialized equipment (Shove et al. 2009; Warde 2014). These four elements also need to be harmonized by users, who have to know how to behave before, during and after the ride. So, as we have seen in section 5, while the company has to create a coordinated and harmonious environment (mainly through formal regulation and a technological platform), users need to collaborate (through common knowledge and feelings) to reach their non-economic goals. The economic issue is not the only goal nor the only variable encouraging the generalized trust that the circuit needs. The theories of practice approach was applied to the study of brand communities (Schau et al. 2009), and it highlighted the fact that users adopt proactive behaviours and are able to create the conditions needed to collectively share goals, knowledge and feelings.

The data that was collected for this research and analysed according to theories of practice show how BlaBlaCar's users tend to evolve as they approach this type of practice. This is done by plotting a trajectory of their relationship with the platform and, above all, with other users (Southerton et al. 2012). We can say that users overcome a state of initial fear and later acquire the skills and knowledge to help them adopt a different approach to the service. Furthermore, they adopt new goals linked to meeting 'new friends'. This kind of approach can be considered as a form of CCC between users that are not 'against something' but rather 'open to' different lifestyles. In this way, BlaBlaCar moves away from a pure market exchange and from critical consumerism. Some users develop new justifications and feelings that still foster mutual collaboration[2].

> I saw that there are definitely different approaches to BlaBlaCar. Those who want only the ride, those who also are looking for friendship. Yes, you can meet new people, but generally, especially as regards the rides I take, which are purely linked to University, there are many opportunities for getting to know other people, other kids… (P 5: Woman 21 years old)

As shown in the previous section, the process of affiliation to the circuit must overcome the fear of the 'stranger'. In the first instance, the aim is economic and individualistic. The goal is achieved through a socio-technical guarantee (i.e. the platform that carries information about other users). Once this period of 'apprenticeship' is complete, the feeling of trust takes on different meanings, which are more collective and collaborative.

> I like BlaBlaCar's ideology not only because it involves an openness to strangers and is therefore a kind of exception if compared to this widely experienced [attitude]. In my opinion, in our communities, in our society, in which everyone has this tendency [to be] a bit 'paranoid': privacy, confidentiality, […] That is, this dimension of fear, often groundless, however, it bothers me. (P 63: Male 22 years old)

In the initial phase, user goals are the same and follow the classic matching of supply and demand. Later on, as practitioners gain experience, goals become shared. Sharing is no longer a coincidence, and it does not only purely relate to a transport service. It involves a willingness to meet new people and to have a more pleasant and enjoyable trip through mutual collaboration. Fun is one of the factors that sets BlaBlaCar apart, beyond the rationale of saving money, and it places the use of BlaBlaCar inside the reciprocal processes of a community of practice.

> I would describe my experience of sharing with a smile, a nice big smile because in the end it was very funny, I met some really nice people, together we shared this journey but it was also a time to know someone and laugh, joke, and sing, and then it was just a new, different experience. If I get a chance I definitely want to do it again. (P 19: Man 28 years old)

---

2 The article uses the concept of trust to describe the feelings and behaviors of people involved in the carpooling practice. In addition, this concept allows researchers to describe how the system can function with strangers as well as other similar economic factors (e.g. monetary practices). The empirical analysis does not distinguish between the concept of trust and reciprocity, which is discussed in the literature of game theory (Cox 2004).

That said, an individual BlaBlaCar ride does not always remain an isolated experience. Many users said that they had met the same people several times – especially in cases of commuting. Some even stated that they had developed close friendships outside of the circuit. The exchange is not offered in terms of profit, but it is related to savings and efficiency. It thus tends to lose its purely economic nature as the user's 'career' progresses.

> No, no, sometimes we find ourselves and we agree, and we travel together and we also stay in contact. In short, it is now a beautiful friendship. (P 13: Male 18 years old)

The data collected do not enable us to state that BlaBlaCar's community is homogeneous. However, it can be argued that, within the different modalities of exchange, users' experiences create and encourage the formation of deeper and stable relationships. These connections allow us to include the BlaBlaCar experience in the realm of CCC because the 'ethos' related to people who frequently use carpooling is a 'kind of community feeling'. Basically, consumers do not choose to share rides for environmental reasons, though they do feel they are part of a 'community of strangers'. Furthermore, word of mouth is the primary channel enabling new users to discover BlaBlaCar. This is supported by the fact that people who have gained some experience with the service strongly recommend using BlaBlaCar to those who have not. It is a platform which not only conveys trust, but also offers the advice and personal experiences of expert users.

> I decided to choose this kind of transport, Blablacar, because [of] the word of mouth of friends and other people I know who had tried [it] and felt fine. Therefore, to reduce costs and to save money I decided to trust myself and to try this new method. (P 26: Woman 20 years old)

> I definitely recommend [it] to all because it is very convenient and then it is something innovative and different, even to try in terms of the relationships that you create. For adults and especially elderly it might obviously be a little 'difficult' in the sense that this type of service uses technology, computer, the Internet, and for an elderly person that may be more difficult; but it could also be interesting for them anyway because it is more convenient than going on a train, going back and forth, so it is recommended to everyone. (P 3: Woman 24 years old)

Based on the collected data, it is the vision of the 'stranger' that is changing. Once a certain degree of experience and familiarity with the service is acquired, people tend to see sharing with a person whom they do not know as a new opportunity that transcends economic circumstances: with BlaBlaCar you have fun, have new experiences and meet new friends. If the stranger is the ideal partner for exchange in advanced monetary economies, then in the sharing economy, could the stranger also be an ideal partner for sharing? BlaBlaCar users are not free riders just because payment and cash are mandatory,[3] but also because part of the real value of the service is the prospect of new relationships.

Conclusion

The article has analysed the BlaBlaCar phenomenon in Italy, using a qualitative approach. Carpooling is a practice that can be included in the realm of the CCC. In fact, even if the practice is market mediated, the experience of carpooling could transform the platform service into a community of strangers. The main purpose of sharing a BlaBlaCar ride is economic, but interestingly, people often feel that this choice is difficult because they must meet and share a private space (the car) with strangers. In fact, an analysis of the data collected reveals that two fundamental codes of interpretation exist.

First, there is economic motivation. Using a car as a collective means of transportation addresses people's desire to save time and money. Secondly, using a car as a shared resource is a personal challenge that leads individuals to trust strangers (Parmiggiani 2013). Environmental protection, one of the possible political considerations of the service, is a secondary motivation. Even if the company supports environmental services through advertising, it does not seem to be users' main goal. These factors suggest that BlaBlaCar users have individualistic orientations that are more closely linked to economic desires – achieving an individualistic benefit – than the need to adopt an environmentally responsible lifestyle. However, these orientations tend to evolve into different practices that have a strong social impact (i. e. the enthusiastic desire to meet new people and to challenge fear).

In the light of what has been said, it is appropriate to think about methodological and sampling choices to identify new directions for further research. While observations were based on people who have used BlaBlaCar at least twice, the opinions of people who have decided to stop using the service should also be analysed. In addition to the factors that support the spread of a feeling of generalised trust, factors that limit this feeling are equally important. A comparison of other sharing practices is also important, even if significant differences between the practices do not lead us to assume that there are specific connections between them. Furthermore, future research might apply a longitudinal logic. This would

3 Payment (in cash) was exchanged face to face when the research took place. BlaBlaCar now mediates all transactions online.

first describe any changes in the approach of respondents over time. Secondly, it would explore how users perceived the introduction of a new form of payment (by researching users who were subsequently free to manage the deal as they preferred, for example, during a period when BlaBlaCar merely brought users together).

In sum, this research is exploratory in nature in that it highlights the dynamic nature of BlaBlaCar users. Individual users enter into a relationship with a community and change that relationship with other users significantly over time. While the platform maintains recursive factors, the use of BlaBlaCar can be described via a trajectory that depends on the level of involvement of the users themselves. The most important result of the research is the clarification of the relationship that practitioners establish with the figure of the 'stranger'. This evolves from being linked to feelings of fear to an opportunity to achieve a better ride – and not only from an economic perspective. For this reason, BlaBlaCar is a community in which users do not perceive people whom they have not met before as strangers. Rather, they think of them as members of the community whom they do not yet know. It is also appropriate, as Zelizer (2010) suggested, to move beyond a pure market model and enter into a social arena within which new relationships are shaped. Accordingly, this may give rise to more stable communities that go beyond the logic of an exclusively economic exchange.

References

Albinsson P.A., & Perera, Y. (2012). Alternative Market Places in the 21st Century: Building Community through Sharing Events. *Journal of Consumer Behavior,* 11: 303–315.

Bardhi, F., & Eckhardt, G.M. (2012). Access-Based Consumption: The Case of Car Sharing. *Journal of Consumer Research,* 39, issue 4, p. 881–98, doi: 10.1086/666376.

Bertolin, A., Beria P., & Filippini, G. (2016). Prime evidenze sul Carpooling in Italia: chi, dove e quando, Rivista di Economia e Politica dei Trasporti, n. 2, art. 1.

Belk, R. (2014). You are what you can access: Sharing and collaborative consumption online. *Journal of Business Research,* 67, 1595–1600.

Botsman, R., & Rodgers, R. (2010). *What's mine is yours: The rise of collaborative consumption.* New York, NY: HarperCollins.

Callon, M. (1998). *The Law of the Market.* Blackwell Publisher, Oxford.

Carfagna, L.B., Dubois, E.A., Fitzmaurice, C., Ouimette, M.Y., Schor, J.B., Willis, M., & Laidley, T. (2014). An emerging eco-habitus: The reconfiguration of high cultural capital practices among ethical consumers. *Journal of Consumer Culture,* 14(2), 158–178.

Chan, N.D. & Shaheen, S.A. (2011). Ridesharing in North America: Past, Present and Future. *Transport Review,* 32(1), 93–112.

Cohen, B., & Kietzmann, J. (2014). Ride on! Mobility business models for the sharing economy. *Organization & Environment*, 27, 279–296 http://oae.sagepub.com/content/27/3/279.short

Cox, J. C. (2004). How to identify trust and reciprocity. *Games and Economic Behavior*, Volume 46, Issue 2, 260–281, http://dx.doi.org/10.1016/S0899-8256(03)00119-2.

Cruz, I. S., & Katz-Gerro, T. (2016). Urban public transport companies and strategies to promote sustainable consumption practices. *Journal of Cleaner Production*, 123, 28–33. doi: http://dx.doi.org/10.1016/j.jclepro.2015.12.007

de Kerckhove, D. (1997). *Connected intelligence: the arrival of the Web society*. Toronto: Somerville House.

Degli Esposti, P. (2015). Peer-to-Peer File Sharing. In *The Wiley Blackwell Encyclopedia of Consumption and Consumer Studies*. London, Wiley Blackwell, pp. 1–2.

Di Minin, A., De Marco, C. E., Marullo, C., Piccaluga, A., Casprini, E., Mahdad, M., & Paraboschi, A. (2016). *Case Studies on Open Innovation in ICT (No. JRC100823)*. Institute for Prospective Technological Studies, Joint Research Centre.

Farajallah, M., Hammond, R. G., & Pénard, T. (2016). *What Drives Pricing Behavior in Peer-to-Peer Markets? Evidence from the Carsharing Platform BlaBlaCar*. Evidence from the Carsharing Platform Blablacar (August 23, 2016).

Fellows, N. T., & Pitfield, D. E. (2000). An economic and operational evaluation of urban car-sharing. *Transportation Research Part D: Transport and Environment*, 5(1), 1–10.

Forno, F., & Garibaldi, R. (2015). Sharing Economy in Travel and Tourism: The case of home-swapping in Italy. *Journal of Quality Assurance in Hospitality & Tourism*, 16(2), 202–220.

Fraiberger, S. P., & Sundararajan, A. (2015). Peer-to-peer rental markets in the sharing economy. NYU Stern School of Business Research Paper.

Furuhata, M., Dessouky, M., Ordóñez, F., Brunet, M. E., Wang, X., & Koenig, S. (2013). Ridesharing: The state-of-the-art and future directions. *Transportation Research Part B: Methodological*, 57, 28–46.

Giddens, A. (1991). *The Consequences of Modernity*. Polity Press; New Ed edition

Hamari, J., Sjöklint, M., & Ukkonen, A. (2016). The sharing economy: Why people participate in collaborative consumption. *Journal of the Association for Information Science and Technology*.

Laamanen, M., Wahlen, S., & Campana, M. (2015). Mobilising collaborative consumption lifestyles: A comparative frame analysis of time banking. *International Journal of Consumer Studies*, 39 (5), 459–467. doi:10.1111/ijcs.12190.

Maturo, A., & Setiffi, F. (2016). The gamification of risk: how health apps foster self-confidence and why this is not enough. *Health, Risk & Society*, 17(7-8), 477–494.

Mongili, A., & G. Pellegrino G. (eds.) (2014). Information infrastructure(s) Boundaries, Ecologies, Multiplicity, Cambridge Scholars Publishing, Newcastle, UK.

Mundler, M., Stocker, A., & Shaheen, S. (2016). Online and App-Based Carpooling in France: Analyzing Users and Practices – A Case Study of BlaBlaCar. In Transportation Research Board 95th Annual Meeting (No. 16-1910).

OECD & International Transport Forum (2015). The Sharing Economy: How shared self-driving cars could change city traffic, available http://www.itf-ecd.org/sites/default/files/docs/15cpb_self-drivingcars.pdf

Pais, I., & Mainieri, M. (2015). Il fenomeno della sharing economy in Italia e nel mondo. *Equilibri*, 1, 11–20, doi: 10.1406/79306.

Pais, I., & Provasi, G. (2015). Sharing Economy: A Step towards the Re-Embeddedness of the Economy? *Stato e mercato*, 3, 347–378, doi: 10.1425/81604.

Paltrinieri, R. (2012). *Felicità responsabile: il consumo oltre la società dei consumi*. FrancoAngeli.

Parmiggiani, P. (2013). Pratiche di consumo, civic engagement, creazione di comunità. *Sociologia Del Lavoro*, (132), 97–112.

Rifkin, J. (2000). The age of access: The new culture of hypercapitalism. Where All of Life is a Paid-For Experience. Tarcher, New York, 33, 40–51.

Schau, H. J., Muñiz Jr, A. M., & Arnould, E. J. (2009). How brand community practices create value. *Journal of marketing*, 73(5), 30–51.

Schor, J. (2014). Debating the sharing economy. Essay published by the Great Transition Initiative, Tellus Institute. http://www.greattransition.org/images/GTI_publications/Schor_Debating_the_Sharing_Economy.pdf

Schor, J. B., & Fitzmaurice, C. J. (2015). Collaborating and connecting: the emergence of the sharing economy. Handbook of research on sustainable consumption, 410.

Scuola Superiore Sant'Anna (2014). Giovani, occupati e tre volte smart: la prima ricerca sui viaggiatori italiani in ride sharing racconta una community che si sposta in modo consapevole. http://www.sssup.it/ist_news.jsp?ID_NEWS=4914&area=199&ID_LINK=10521&GTemplate=ist_home.jsp

Secondulfo, D. (2016). *Il mondo di seconda mano. Sociologia dell'usato e del riuso*. FrancoAngeli, Milano.

Secondulfo, D. (2012). The three consumer profiles in Italy. *Italian Sociological Review*, 2(3), 125.

Setiffi, F., & Lazzer, G. P. (2016). Di madre in madre. L'immaginario dei negozi di seconda mano per bambini. In D. Secondulfo (Ed.), *Il mondo di seconda mano. Sociologia dell'usato e del riuso*. FrancoAngeli, Milano.

Shove, E., Trentmann, F., & Wilk, R. (2009). *Time, Consumption and Everyday Life: Practice, Materiality and Culture*. Oxford: Berg.

Simmel, G. (1900). *The Philosophy of Money*. London: Routledge (2004).

Southerton, D., Olsen, W., Warde, A., & Cheng, S. L. (2012). Practices and trajectories: A comparative analysis of reading in France, Norway, the Netherlands, the UK and the USA. *Journal of Consumer Culture*, 12(3), 237–262.

Teal, R. F. (1987). Carpooling: who, how and why. *Transportation Research Part A: General*, 21(3), 203–214.

Vincent-Geslin, S. (2008). Les "altermobilités": analyse sociologique d'usages de déplacements alternatifs à la voiture individuelle: des pratiques en émergence? (Doctoral dissertation, Paris 5).

Wahlen, S., & Laamanen, M. (2017). Collaborative Consumption and Sharing Economies. In M. Keller, B. Halkier, T.-A. Wilska & M. Truninger (eds.), *Routledge Handbook on Consumption*. Oxford, UK: Routledge.

Warde, A. (2014). After taste: Culture, consumption and theories of practice. *Journal of Consumer Culture*, 1469540514547828.

Yates, L. (2016). Sharing, households and sustainable consumption. *Journal of Consumer Culture*, 22, doi: 10.1177/1469540516668229

Zelizer, V.A. (2010). *Economic lives: How culture shapes the economy*. Princeton University Press.

Zelizer, V.A. (2007). *The Purchase of Intimacy*. Princeton, NJ: Princeton University Press.

The sharing economy and young people: an exploratory research project

Ariela Mortara and Geraldina Roberti

Abstract

The chapter presents the results of an explorative survey aimed at understanding sharing economy (SE) related practices among Italian university students. Perception, motivations and practices are explored through a web based questionnaire. Among the results, the issue of trust emerges as relevant, however the lack of trust is not the main cause of the limited use of SE practices. Indeed, students are keen to verify the reliability of people and platforms through online reviews fostering the collaborative platforms reputation.

Zusammenfassung

Das Kapitel stellt die Ergebnisse einer explorativen Umfrage vor, die darauf abzielt die Praktiken im Bereich der Sharing Economy unter italienischen Studenten zu verstehen. Wahrnehmung, Motivation und Praktiken werden durch einen webbasierten Fragebogen erforscht. Unter den Ergebnissen stellt sich das Thema Vertrauen als relevant heraus, jedoch ist der Mangel an Vertrauen nicht die Hauptursache für den begrenzten Einsatz von kollaborativen Konsumpraktiken. In der Tat sind Studierende in der Umfrage daran interessiert, die Zuverlässigkeit von Menschen und Plattformen durch Online-Bewertungen zu überprüfen, und dadurch den Ruf der Plattform des kollaborativen Konsums zu fördern.

Keywords

sharing economy, university students, trust

I. Cruz et al. (Hrsg.), *Contemporary Collaborative Consumption*,
Kritische Verbraucherforschung, https://doi.org/10.1007/978-3-658-21346-6_6

Introduction[1]

In recent years, both sharing and collaborative consumption (CC) have grown in popularity[2]. Researchers shifted their focus from issues related to ownership of goods to their use. Owyang and Samuel (2015) point out that the collaborative economy is the biggest shift in the business landscape since the advent of the Internet itself. It is a key change, since it affects the dynamics of postmodern identity construction. According to several scholars (Bauman 2007; Belk 1988), postmodern society allows social actors to act more creatively and to change mechanisms of self-definition. As a result, individuals have learned to use the cultural significance of objects both to build their own identity and to realize an original biographical project (Beck 1992). Moreover, the *possession* of goods/products has become a fundamental factor enabling social actors to incorporate the symbolic meaning of objects in their own *extended self* (Belk 1988). However, recent changes in society, together with the intensifying of the economic crisis, have partially shifted consumers' attention to the use of an asset, rather than its ownership.

Consumers are engaging more and more frequently in sharing activities such as renting, lending, trading, bartering and swapping of goods, services, space or transportation solutions (Möhlmann 2015). According to Sundararajan (2013, p. 2), "if you don't need to own the assets you use, not only do you spend smarter, but your product variety and quality options expand". Owning an asset is no longer mandatory – for a growing number of consumers, using a product or a service when needed is more than enough. In this way, consumers enjoy the freedom of being able to change their mind whenever they want. Web 2.0 boosted the consolidation of these changes by fostering a multiplicity of digital platforms able to accommodate a growing number of exchanges and interactions. Through these online-based communities, consumers have been able to coordinate their activities, developing horizontal and egalitarian models of interaction and using these weak ties as social capital dynamically to construct their social networks of consumption (Rosen, La Fontaine and Hendrickson 2011).

Regardless of the role played by such emerging socio-technical systems, however, the underlying paradigm of social action is the most significant change, namely the establishment of a *sharing ethos* to which a growing number of indi-

1 The chapter is the joint work of its two authors. However, following standard academic practice, it should be mentioned that Geraldina Roberti wrote paragraphs 1 and 2 and Ariela Mortara wrote paragraphs 3 and 4. The rest of the paper has been written jointly.

2 The definition of collaborative consumption utilized in this chapter is based on Hamari, Sjöklint and Ukkonen (2015), who specify it as "the peer-to-peer-based activity of obtaining, giving or sharing access to goods and services, coordinated through community-based online services" (p. 3).

viduals subscribe[3]. Several factors foster the gradual spread of various forms of the sharing economy (and collaborative consumption) such as 1) consumers' increased attention to the environmental impact of their choices; 2) a more cautious attitude to spending (partly as a result of the global financial crisis that hit Western societies); 3) a widespread pro-social sensitivity; 4) the role played by new digital technologies.

Indeed, concern with ecological impact positively influences people's attitude to collaborative consumption. As Möhlmann has written (2015, pp. 195–196), "sharing solutions are generally considered to have a positive environmental impact compared with non-sharing solutions because the pooling of material goods leads to the increased intensity in the usage of one single product entity. The material required for each episode is reduced, waste is avoided, and overproduction is countered […]". Often an individually owned product with limited usage is replaced with a shared service, maximizing its utility. The second element affecting the spread of sharing practices is represented by the economic difficulties that individuals have gone through in recent years. Opting for sustainability and reduction in consumption saves costs, but it also means criticizing overconsumption and the dominant capitalistic model (Böckmann 2013). Moreover, according to several authors (e.g. Schor 2014), sharing what you own can result in a supplementary income, since these technological platforms allow people to earn money in ways not previously available. Motivations may thus range from saving money to earning money, since cost consciousness is moving people away from past forms of consumerism toward sharing and cooperation (Botsman and Rogers 2011).

As far as sociability and interactions are concerned, Nelson and Rademacher (2009) have highlighted that community memberships or the aspiration to be part of a community are crucial to practicing collaborative consumption. Indeed, sharing activities convey social and relational meanings, thus ensuring that participants pursue a common goal (Albinsson and Perera 2012). This feeling strongly intertwines with common responsibility to the environment and collective well-being, giving life to practices of *mindful consumption*. Shet, Sethia and Srinivas (2011) define such practices as caring and being conscious, in thought and behaviour, of the implications and consequences of individual and collective consumption. Some of these consumers are, in fact, quite critical of overconsumption. They engage in sharing activities as an intentional political act and are willing to construct an alternative to the usual market (Albisson and Perera 2012).

---

3 According to Belk (2014b, p. 7), "in the economy of things, you are identified by what you own – your land, house, car. In the economy of ideas that the web is creating, you are what you share".

Finally, the technological infrastructure plays a significant role in facilitating coordination and interaction among consumers. Indeed, the user-friendliness of sharing platforms (and the management of online payments) has become a key driver for the development of the different types of collaborative consumption[4]. The massive expansion of laptop computers, tablets, and other handheld electronic devices allows users to be constantly connected and promotes exchanges and business transactions. According to Campbell Mithun's 2012 research, rational or emotional reasons drive consumers to engage in various activities of the sharing economy. Saving and practicality are the rational reasons, while, among the emotional benefits, "generosity to others and myself" is prominent, because it helps people feel altruistic and part of a community. As Schmitt (1999) has written, rational benefits reflect product usefulness and quality for consumers, while emotional benefits concern human emotions. Especially for younger consumers, emotional motivations also include the desire to feel smart and fashionable. According to Möhlmann (2015), young people could engage in collaborative consumption practices to follow an emerging trend and to use innovative and fashionable products and services[5].

Sharing specific services can also be a culturally and socially stimulating experience. As regards hospitality services, e.g. Airbnb or CouchSurfing, promoting cultural experience and understanding among people of a different background emerge as leading motivations for engaging users (Lauterbach, Truong, Shah, Adamic and Arbor 2009). To recap, interest in innovation, attention to social relations, but also to environmental sustainability, ethics, and vision, which are embedded in the sharing of goods and services, are critical factors in supporting the spread of the sharing economy. These trends are so established that they may outgrow the economic crisis.

As to Italy, recent research (Ipsos 2014) reveals that 75 % of the population aged between 18 and 64 have some notion of the sharing economy, with peak awareness among the youngest (18–34). According to respondents these practices are one of the most effective (and handy) responses to the crisis. 11 % of the sample state they are members of such platforms and about one third of the sample (31 %) state they are interested in consumption practices. An even more recent study (Capeci 2015) – on a sample of 1,000 individuals representing the Italian population – confirms that young people aged between 18 and 34 represent 46 % of users of some sharing service (mostly related to mobility or accommodation services). As some

4 According to Schor and Fitzmaurice (2015) it is a real *Technophilia* that distinguishes a substantial part of these connecting consumers.

5 Indeed, in a *Harvard Business Review* paper, Lubin and Esty (2010) defined sustainability as a "business megatrend".

researchers have highlighted (Pais and Mainieri 2015), in Italy the supply of sharing services is still greater than the demand. Except for mobility platforms, the majority of Italian sites have no more than ten thousand users. The limited usage of new communication technologies is one of the biggest barriers to the adoption of sharing services. Indeed, the weight of the digital divide is still significant, especially among the most adult users[6]. Furthermore, the existence of a small number of users restricts significantly the amount of shared goods, reducing the likelihood of finding the desired product/service on the various platforms. As Botsman and Rogers (2011) have stressed, collaborative consumption needs a *critical mass* of users to become self-sustaining. Indeed, to compete with conventional shopping there must be enough choice for all consumers. As the number of people increases, the likelihood of users being dissatisfied decreases.

This chapter presents the results of an exploratory survey aimed at understanding practices related to the sharing economy among Italian university students aged between 18 and 24 years. Since young people seem to be critical for such practices, this age group is the core target for future development of the sharing economy. A deeper understanding of perception, motivations, and actual practices of sharing economy services should highlight future trends in collaborative consumption. The main aim of this research is to understand the true motivations behind sharing economy activities throughout Italy. Practices, perception, and motivations have been explored through a web-based questionnaire (Sills and Song 2002), and the sample gathered using a snowball procedure (Handcock and Gile 2011).

It is all about trust

The rise of collaborative consumption as a possible new paradigm of relations between individuals, trading, and society often revolves around the issue of trust. As Schor and Fitzmaurice (2015) have highlighted, trust and reputation are central questions in the literature on the sharing economy. Indeed, according to Belk's (2010) extended theoretical review of the sharing concept and of its relevance for consumer research, sharing creates solidarity and "[…] goes hand in hand with trust and bonding" (p. 717).

6 As of December 2015, Audiweb data estimated that in Italy just 24.6 % of people in the 55–74 age range (3.5 million unique users) access the Internet on an average day through all detected devices (pc, Smartphone, tablet, TV) http://www.audiweb.it/news/total-digital-audience-del-mese-di-dicembre-2015/.

Indeed, concluding a transaction with a stranger through the different sharing platforms involves the construction of a fiduciary relationship. User reputation and peer-to-peer evaluations are the pillars of this relationship. Some researchers (among them Hardin 2002 and Stolle 2002) distinguish between *particularized* and *generalized trust* (or social trust), which is immediately applicable to collaborative consumption. Indeed, attitudes of generalized trust extend beyond the borders of face-to-face relationships, engaging people who are not personally known but with which a specific type of interaction has been developed[7] – such as swapping or renting. Bicchieri, Duffy and Tolle (2004) mention that "impersonal trust" emerges when interactions among players last for extended periods. When impersonal trust occurs, the recurrence of non-anonymous interactions ensures some kind of informal *social control,* thus safeguarding individuals from the risks associated with not knowing their interlocutors (Bicchieri et al. 2004). According to the same researchers (Bicchieri et al. 2004), a network of *repeated* exchange relationships is a source of trust, since people will prefer to negotiate with individuals with good reputations. How then can users protect themselves from the risks of "one-shot" interactions where there is no pre-existing social relationship? Furthermore, if trust also relies on the content and extent of the interaction (Rohe 2004), how do users of sharing economy products/services trust strangers with whom they have never dealt before?

Moreover, according to Stewart's researches "a hypertext link from one Web site to another increased the extent to which the linked organizations were perceived to have a business relationship and be similar, and these perceptions had a positive influence on trusting beliefs regarding the linked site" (2003, p. 1).

As trust is a crucial factor for the success of an online sharing platform (Beldad, de Jong and Steehouder 2010; Finley 2013), sites such as Couchsurfing or Airbnb have developed a sophisticated reputation system that offers references and immediate feedback on the reliability of each user of the platform. As Finley (2013) notes in connection with Airbnb, a similar references model transforms the individual's reputation into a "collective measure of trustworthiness calculated from other community members' referrals and ratings" (p. 18). On CouchSurfing "users can only vouch for others if they have at least three vouches themselves. […] The purpose of the vouching system is to give members additional information to aid in judging whether another member is trustworthy for 'hosting' a visitor or 'surfing' on a host's couch" (Lauterbach, Truong, Adamic and Arbor 2009, p. 346). Ultimately, these tools are a valuable source of information for consumers, as they enable them to acquire news and opinions on the person with whom they will interact. Indeed, the Internet is a real "cornucopia" of shared information available

7 Indeed, Schor (2014) writes about "stranger sharing" (p. 7).

to all users (Belk 2010), and the added verification features have a positive effect on the trustworthiness of the individuals. This "the more information, the better" philosophy (Finley 2013) allows consumers to minimize the online transactions risk and transforms the reputation of collaborative platforms users *into a new kind of social capital.*

Similar dynamics may undermine the claims of a scholar like Putman (2000), who highlighted the progressive decline of civic engagement in American society and the weakening of confidence in communities. Nowadays, community members' roles are paramount, since other users share specific consumption with them and their comments are the best guarantee of the quality of the actual product or service. In horizontally organized structures, such as those offering collaborative consumption, all network nodes are at the same level, thus ensuring direct and equal information exchange among users and an increase in the confidence level.

The chapter structure is as follows. The following paragraph outlines the methodology, the characteristics of the sample and some of the main survey findings. The next section discusses findings relating to trust. The final section offers some concluding remarks, and points to areas for further research. This brief literature review provides the theoretical foundation for the subsequent empirical research.

Methodology, sample, and descriptive statistics

The authors carried out online quantitative research through the Google Drive platform. Questionnaires were collected over two months (from April to May 2016). To satisfy authors' aims, the questionnaire 1) collects information about possible changes in consumption practices due to the economic crisis; 2) assesses the level of knowledge concerning the sharing economy; 3) verifies the extent to which the most common sharing economy practices have been disseminated; 4) enquires as to the motivation leading to the decision to use collaborative economy platforms or not; 5) collects socio-demographic information on participants and their families. In constructing the questionnaire, the authors drew on recent studies by Capeci (2015) and Ispos (2014), which survey the spread of sharing economy practices among Italian consumers. The data collected were analysed using Microsoft Excel and SPSS.

The convenience sample (Coomber 1997) of 283 respondents was gathered while presenting the research during several classes in four different Italian universities, namely the University of Trento (Unitn), IULM University of Milan (IULM), Sapienza University of Rome (Sapienza), and University of L'Aquila (Univaq). Most of the students are attending Communications faculties. The sample was collected using a snowball procedure (Handcock and Gile 2011).

Table 1 Student attendance at different universities

Unitn	13.1 %
IULM	43.8 %
Sapienza	37.8 %
Univaq	5.3 %
Total	100.0 %

Universe: all respondents.

Table 2 Student attendance at universities (degree level)

	First degree	Master's degree	Total
Unitn	0 %	100 %	100 %
IULM	74 %	26 %	100 %
Sapienza	86 %	14 %	100 %
Univaq	93 %	7 %	100 %

Universe: all respondents.

The four universities represent the North (Trento and Milan) and Centre (Rome and L'Aquila) of the country. The choice of these two geographical areas enabled researchers to compare attitudes toward collaborative consumption of young people living in parts of the country with different levels of sociocultural development. Indeed, in Italy, according to numerous sources (Scarpellini 2011; Istat 2014 and Bella 1988), consumption indicators point to inequality between the North, the Centre and the South of the country. Distribution of income and wealth penalize the central and southern regions, as highlighted by the difference in the values of equity between North and South[8] (a ratio of more than 2 to 1).

Most respondents (76 %) are female and, as far as age is concerned, the two largest groups are those born between 1995–1996 (48 %) and 1991–1993 (32.2 %). Overall, 70 % of students are taking first degrees and 30 % the master's degree. Tables 1 and 2 summarise the composition of the sample by university and degree level.

8 On this topic, see http://agriregionieuropa.univpm.it/it/content/article/31/36/la-distribuzione-del-benessere-italia-diversita-tra-famiglie-urbane-e-rurali.

Thinking about the sharing economy

Since, according to previous research (Belk 2014a; Ipsos 2014; Capeci, 2015), the economic crisis of recent years encouraged the spread of sharing economy practices, the first questions in the questionnaire address this topic. 70 % of those sampled declare that the economic crisis has changed their life-style and some consumption practices. Figure 1 shows that students have mostly cut down on some types of superfluous consumption, most likely those they have to pay for themselves.

Since the sharing economy (SE) is a relatively new phenomenon, the next question assesses the knowledge of this practice among students. 60.1 % of the sample say they have heard of it. More interesting is the breakdown of the sample by university. Table 3 shows that almost 66 % of students with knowledge of the SE come from universities based in the North of Italy (Trento and Milan), although these two universities together represent approximately 57 % of the total sample. This gap highlights the typical socio-cultural differences in the various areas of the country, hosting a North traditionally more advanced and evolved and a South-Centre (Malanima and Daniele 2007) struggling to keep up with European standards – even in the field of modern consumption.

The direct comparison between the two main cities of the country, Milan and Rome, is equally significant. Despite a similar consistency in the sample, students

Figure 1 Which goods and services do you use differently compared to the past?

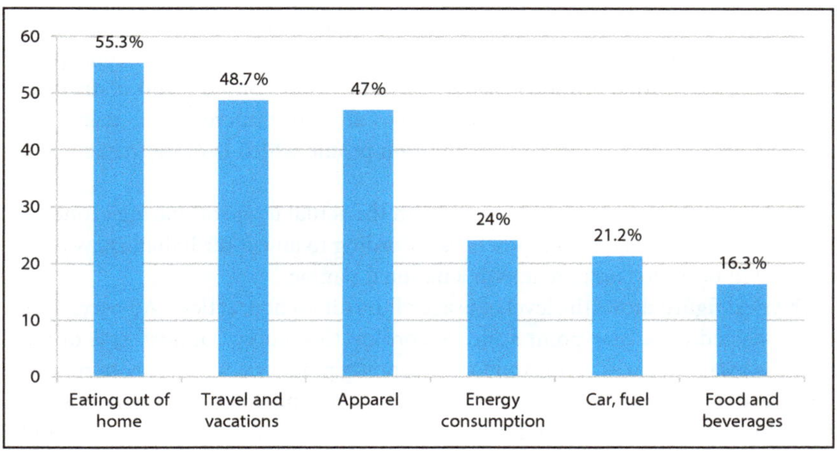

Universe: all respondents. Multiple answers.

Table 3 Knowledge of the Sharing Economy by University

		Which University are you at?				Total
		Unitn	IULM	Sapienza	Univaq	
Have you ever heard about the Sharing Economy (or the collaborative economy)?	YES	16.5 %	49.4 %	29.4 %	4.7 %	100 %
	% of Total 60.1 %					
	NO	8.0 %	35.4 %	50.4 %	6.2 %	100 %
	% of Total 39.9 %					
Total Sample		13.1 %	43.8 %	37.8 %	5.3 %	100 %

Universe: all respondents.

in Milan display a substantially higher level (49.4 % vs. 29.4 %) of knowledge of the sharing economy (Table 3). One of the reasons for this gap may be found in the different awareness of bike sharing – the second-best known sharing economy practice (Figure 3 below). Also, thanks to the different urban landscape of the two cities, bike sharing is much more common in Milan than in Rome.

The following question asks students to associate the SE with one of the five suggested concepts. Almost half of the interviewees chose the idea of sharing of goods. Interestingly, association with "sharing a service" was chosen by less than 20 % of the sample. Only 5.4 % of respondents associate the SE with saving, even if that is one of the leading motivations for such practices (see Fig. 4 below). Figure 2 ranks all the associations.

The intersection of this outcome and respondents' gender (Table 4) highlights the fact that for men, besides sharing, saving is equally linked to the concept of sharing economy. Men are thus also interested in the economic and instrumental aspects of collaborative practices. Females are more interested in sharing issues. The data point to a female consumption profile which is more sociable and interactive.

The next questionnaire section deals with the actual usage of sharing economy practices. The authors selected practices according to an earlier Italian survey (Ipsos 2014) in order to compare it with a national sample.

The next figure shows the level of usage of the different practices. Answers have been collected on a four-point scale. According to Figure 3, almost 75 % of the sample is aware of car-sharing and local shopping practices, while 64 % of respondents have never heard of couchsurfing. This data confirms the results obtained by Rosen et al. (2011), which indicate that the mean age of couchsurfers is 27, and these individuals are older than the average age of the analysed sample. Moreover,

Figure 2 Thinking about the sharing economy, what comes to mind for you?

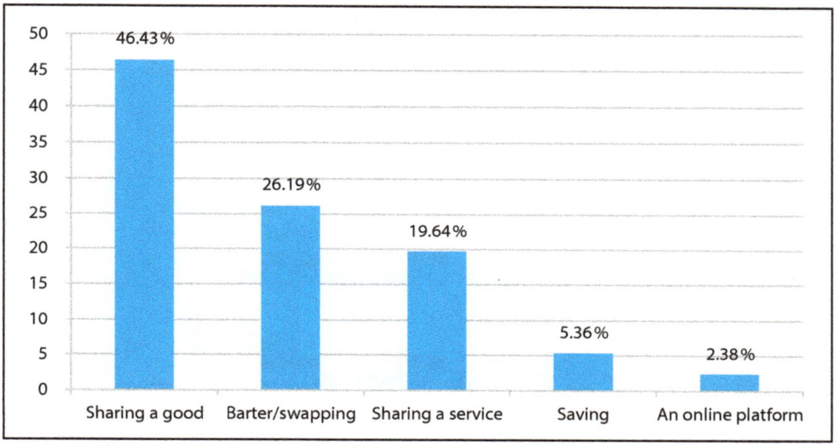

Universe: people knowing the SE (n = 168).

Table 4 Perception of the SE by gender

		Sex	
		M	F
Thinking about the sharing economy what comes to mind for you?	Sharing a good	11.90 %	22.22 %
	Barter/swapping	4.76 %	5.56 %
	Sharing a service	40.48 %	48.41 %
	Saving	40.48 %	21.43 %
	An online platform	2.38 %	2.38 %
Total Sample		100.0 %	100.0 %
People knowing the SE		42	126

Universe: people knowing the SE (n = 168).

Figure 3 For each of the following practices, can you tell me if you know them and how often you use them?

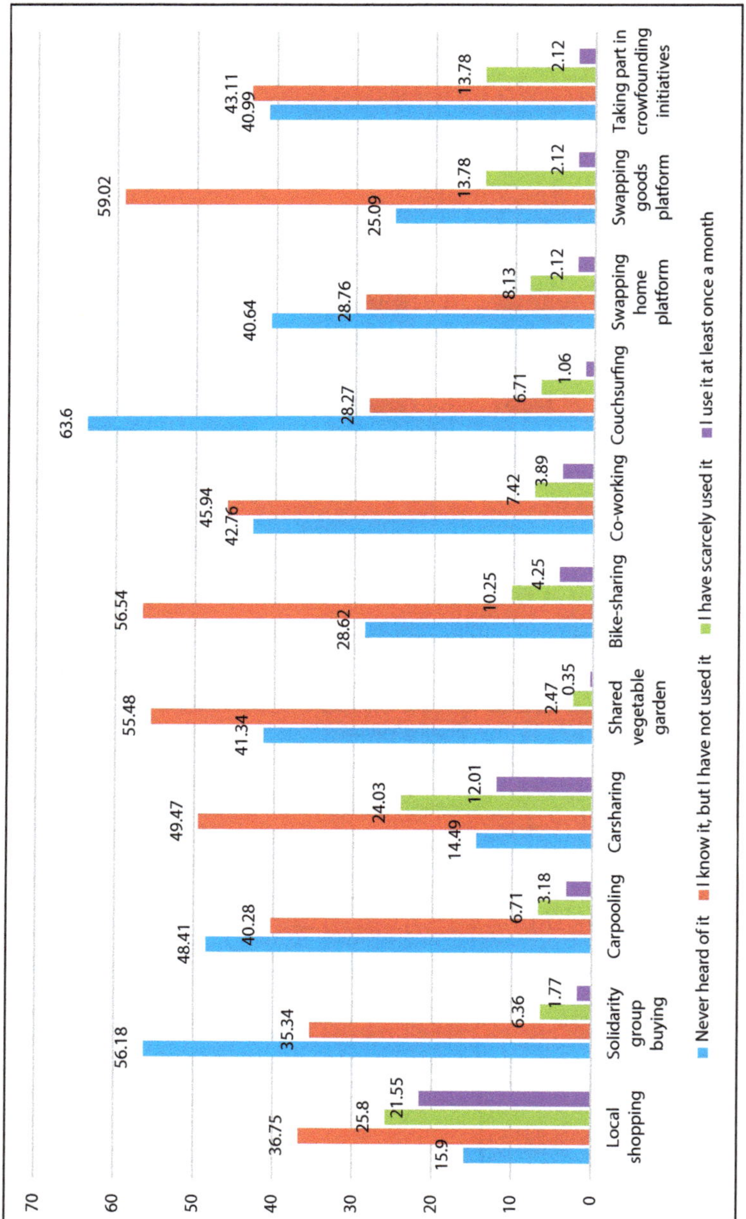

Universe: users of a single practice.

couchsurfing is all about sharing a couch, and, university students are unlikely to own a couch to share, since they live with their family or share an apartment with other students. Furthermore, almost half of the sample is very young (19–20 years old), hence less likely to travel alone (without family or friends), and is probably looking for more conventional accommodation.

Students shop locally (21.55 %) and share cars at least once a month. Car sharing is a distant second, with a percentage of 12.01 %. According to Eurisko (2016), in Italy car sharing, along with peer-to-peer accommodation, represents the breakthrough phase of the SE lifecycle, while peer to peer landing and crowdfunding are still a niche. These data partially confirm those provided by Censis (2015), according to which 8.4 % of Millennials use bike sharing and car sharing (compared to the 4.1 % of the next generational cohort of people between 35–64 years old).

Surprisingly, since the BlaBlacar phenomenon gained very high media visibility (Cowan 2015) almost half of the sample does not know carpooling (or ridesharing). Again, the young age of respondents could be the reason why they do not use ridesharing services. The generational cohort also influences the actual use of carpooling and carsharing, which is significantly higher among master's degree students (1991–1993 cohorts) compared to first-degree students (1995–1996 cohorts). As for carsharing, among those who said they had occasionally used the service, over 42 % are doing a master's degree (while the total weight of this cohort in the sample is only 30 %).

As regards local shopping, the more widespread practice, table 5 points to a significant increase within the female universe (more than 10 percentage points) in comparison with the male part of the sample. A possible explanation for such differences might be the role of caregivers, still mostly performed by women. Indeed, according to several researchers (e.g. Tanturri 2011 and de Singly 1994), in Italy an obvious gender asymmetry still affects the division of family tasks. Women are mostly in charge of domestic chores, in a context of particularly stiff and traditional gender roles. Thus, as Cairns, Johnston, and MacKendrick (2013) have emphasized, women's commitment to the family implies a specific focus on household consumption and purchases. In fact, women are also much more concerned with

Table 5 Local shopping users by gender

	Non-users of Local shopping	Users of Local shopping	Total
Male	60.3 %	39.7 %	100 %
Female	50.2 %	49.8 %	100 %

Universe: all respondents.

Table 6 Local shopping users by university

	Non-users of Local shopping	Users of Local shopping	Total
Unitn	45.9 %	54.1 %	100 %
IULM	52.4 %	47.6 %	100 %
Sapienza	57.0 %	43.0 %	100 %
Univaq	40.0 %	60.0 %	100 %
Total	52.7 %	47.3 %	100 %

Universe: all respondents.

the quality of the products they provide for the family. A number of studies, in different national contexts, point to the role of women as the strongest supporters of organic food and the use of non-industrial raw materials. Local shopping guarantees product quality, which is so important for the female part of the sample.

Furthermore, according to table 6, local shopping is most widespread among students enrolled in the universities of Trento and L'Aquila, two much smaller cities compared to Rome and Milan. Indeed, more than half of students of the first two universities (respectively 54.1 % and 60 % of respondents) said they used forms of local shopping. It can be argued that territorial proximity with the production site of goods (especially groceries) ensures ease of access, thus encouraging the spread of the practice (Handy and Clifton 2001).

The next question addresses the motivations for such practices which, according to previous research, have different roots. Capeci's data (2015) underline the economic crisis, thus the need to save money, as a driving force. Likewise Owyang and Samuel (2015) point out that sharing is saving and, mostly for younger customers, saving money is one of the top drivers of the collaborative economy. Among others, Harrison, Newholm and Shaw (2005) highlight the increasing role of ethical concerns as generally driving consumption practices. According to Möhlmann (2015), engaging in collaborative consumption practices could be fashionable for young people; and, obviously, defining one's identity (Paterson 2009) through consumption is always a core issue. Figure 4 shows the motivations leading to the different practices.

The need to save drives young people to use car services and swapping platforms. Bike-sharing and couchsurfing provide new experiences, although very few interviewees have actually used them (see Figure 3). Surprisingly, "feeling smart" does not appear as a driver for SE practices and this could point to such trends no longer being a fad. Indeed, for young people the innovative potential of collabora-

Figure 4 Thinking about the activities/services that you have already used, can you tell us what is the leading motivation that made you choose them?

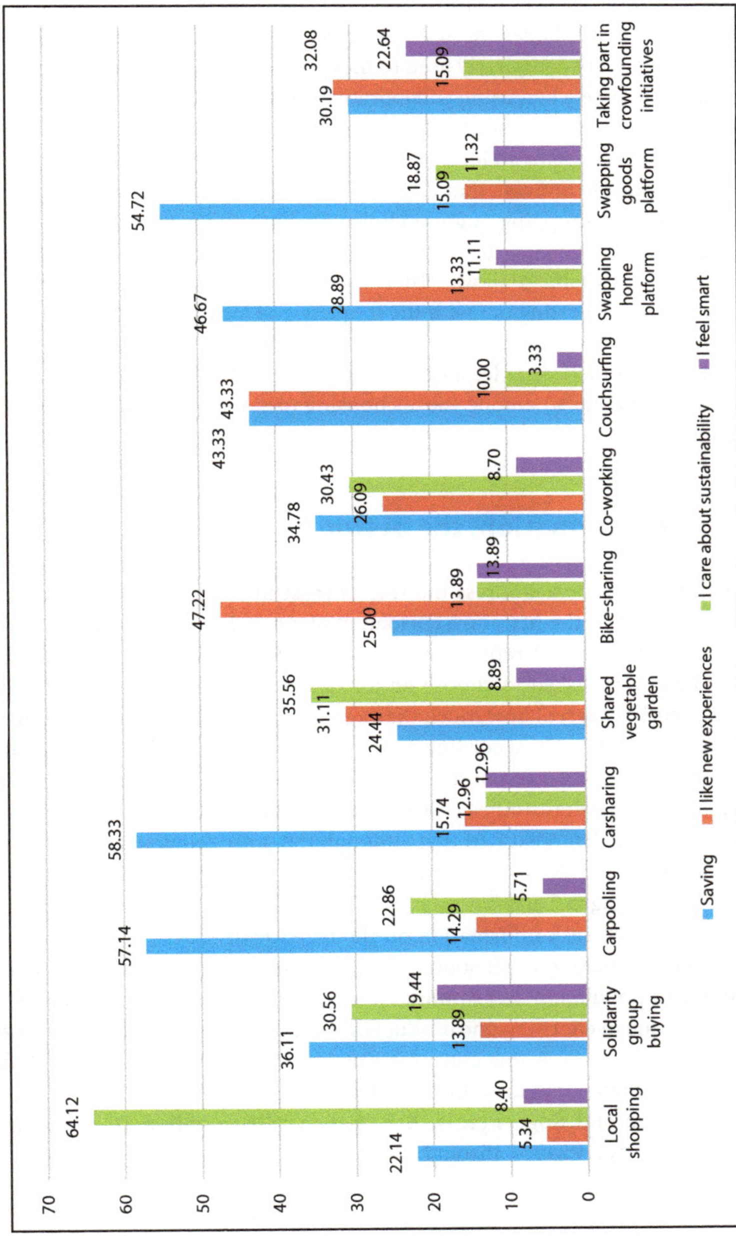

Universe: users of a single practice.

tive consumption services is already wearing thin. The innovative nature of these practices no longer shapes the self-perception of very young people. Older consumers, who are not completely used to these practices, may still see themselves as cool and up to date when using collaborative consumption services.

Furthermore, caring about sustainability and saving are the leading motivations for people engaging in local shopping. Previous research (Brown et al. 2009; Coley et al. 2009) confirms that caring for the environment is a leading motive in choosing local shopping, a good choice for both the environment and the local community. According to the sample, the need to save represents a second strong motive, as table 7 shows. This table is obtained by cross-referencing reduced spending on eating out with the frequency of local shopping. Table 7 shows that, among those who do not report lower spending, 41.2 % claim to have adopted local shopping practices at least occasionally. This percentage rises by more than 10 points (up to 51.9 %) among those choosing to save by limiting their spending on eating out.

The more information, the better

The next two questions address the role of trust in choosing these practices. Indeed, according to the literature (Beldad et al. 2010; Finley 2013; Schor and Fitzmaurice 2015) trust and reputation are key issues in this kind of consumption. Respondents were asked to choose on a five-point Likert scale (1 = strongly disagree; 5 = entirely agree) how much they agreed with various statements reflecting the different positions[9]. According to Figure 5, before engaging in collaborative consumption practices, respondents evaluate how reliable they are by looking at online reviews. Several studies (David and Pinch 2005; Mayzlin, Dover and Chevalier 2014) have demonstrated that although the online review system cannot always be trusted, its role is still of paramount importance. 75 % of respondents (the sum of "entirely agree" and "mostly agree") find out about the people offering SE services, and 63 % seek information on the trustworthiness of the platform. This attitude is confirmed by other sources (e. g. Eurisko 2016), according to which 57 % of the Italian population seeks online information, and trust represents the most important issue in choosing one product or service over another. Data obtained by Wagner et al. (2015) show that consumers use contacts with other users of different platforms to obtain assurances of the reliability of SE services.

As Botsman and Rogers (2011, XIV) have written, "online exchanges mimic the close ties once formed through face-to-face exchanges in villages, but on a

9 For the complete items lists, see tables 12 and 13 of Appendix.

Table 7 Reducing spending on eating out by frequency of local shopping

		Local shopping				Total
		Never heard of it	I know it, but I have not used	I have scarcely used it	I use it at least once a month	
Spending on eating out	NO	19.8 %	38.9 %	21.4 %	19.8 %	100.0 %
	% of Total	8.9 %	17.4 %	9.6 %	8.9 %	44.7 %
	YES	12.8 %	35.3 %	29.5 %	22.4 %	100.0 %
	% of Total	7.1 %	19.5 %	16.3 %	12.4 %	55.3 %
Total sample		16.0 %	36.9 %	25.9 %	21.3 %	100.0 %

Universe: all respondents.

Figure 5 Thinking about when you are using one of these new activities/services, express your level of agreement with each of the following statements

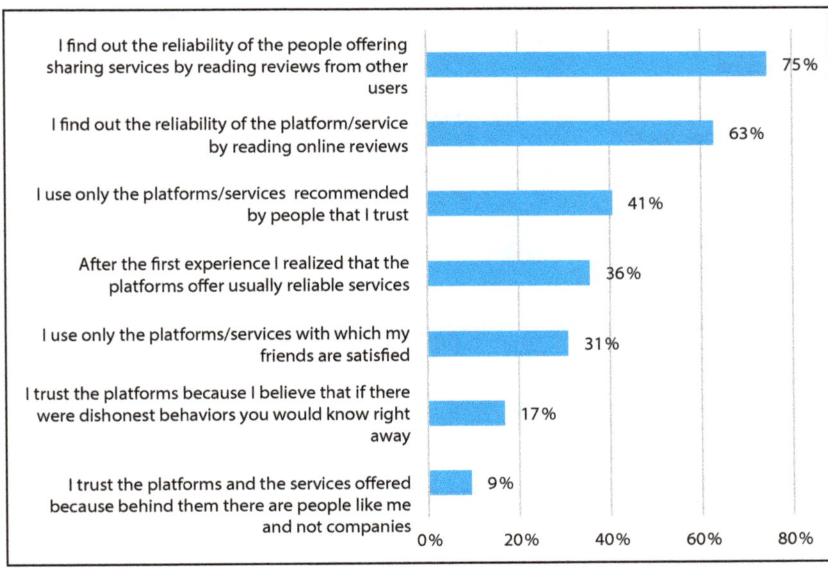

Universe: users of at least one practice. Sum of "entirely agree" and "mostly agree".

Figure 6 I find out about the reliability of the platform/service by reading online reviews

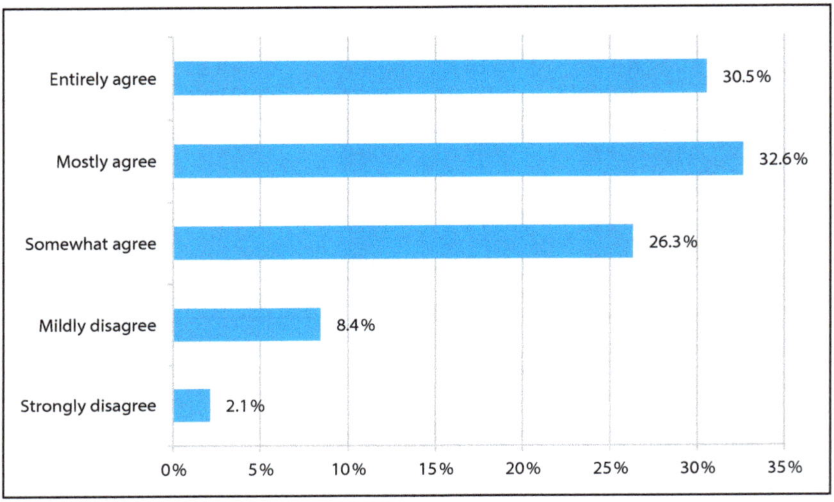

Universe: users of car-sharing services. Universe: uses of car-sharing services.

Figure 7 I find out about the reliability of people offering sharing services by reading reviews from other users

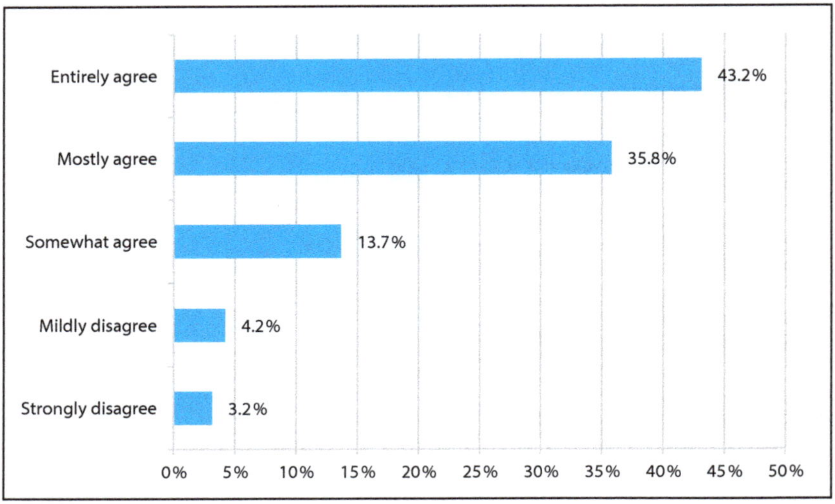

Universe: users of car-sharing services. Universe: uses of car-sharing services.

much larger and unconfined scale. In other words, technology is reinventing old forms of trust". Online reviewers are trusted more than "trusted people" (41 %) or "friends" (31 %), which seems strange, since the latter represent known people, while online reviews are anonymous. Just 9 % of the sample agrees that the human factor should be trusted in itself.

As a further confirmation, the paramount role of the web as a source of information and reassurance emerges from the precautionary strategies of car-sharing users (the second most used practice after local shopping). In order to protect themselves against the risks of using such services, more than 60 % of the students (sum of "entirely agree" and "mostly agree") rely on peer-to-peer reviewers (figures 6 and 7) both for the platform and for the people offering the services.

In some ways, a positive rating from a large number of people becomes equivalent to a first-hand rating from those whom users have effectively met. Personal/online relationships and social capital are again at the heart of the exchange, helping consumers to make better decisions. As Wagner et al. (2015) have highlighted, connecting people and increasing social interaction, such as in co-creation spaces, are very important aspects of the SE value proposition. Indeed, collective consumption differs from other, more efficient digital markets. The Sharing Economy is in many ways a return to the intense social interaction that historically characterizes physical marketplaces.

Moreover, data analysis shows that trust components are different according to gender. Figure 8 shows that the items in which there are the greatest differences between the genders are the second and the third items – the answers "I find out about the reliability of the platform/service by reading online reviews" and "I use only the platforms/services recommended by people that I trust" (sum of "entirely agree" and "mostly agree"). For both items, girls' answers show the highest percentage. Compared to male part of the sample, the gap is nearly 20 percentage points for the first item and about 10 points for the second. To explain this significant difference, it can be argued that the female part of the sample is more involved in the relational aspect of consumption practices. In other words girls, while using specific SE practices, trust other users' opinions more than boys do. They also discuss the reliability of the different sharing platforms with their friends.

The difference between male and female is confirmed by the following variance analysis (Table 8). Females seem to be more compact in their choices. Specifically, male answers are more scattered around "The importance of reading online reviews about the people offering SE services".

The next question investigates the reasons why interviewees have not used SE products or services. Again, respondents were asked to choose on a five-point Likert scale (1 = strongly disagree; 5 = entirely agree) how far they agreed with some items examining their different motivations.

Figure 8 Thinking about when you are using one of these new activities/services, express your level of agreement with each of the following statements (according to gender)

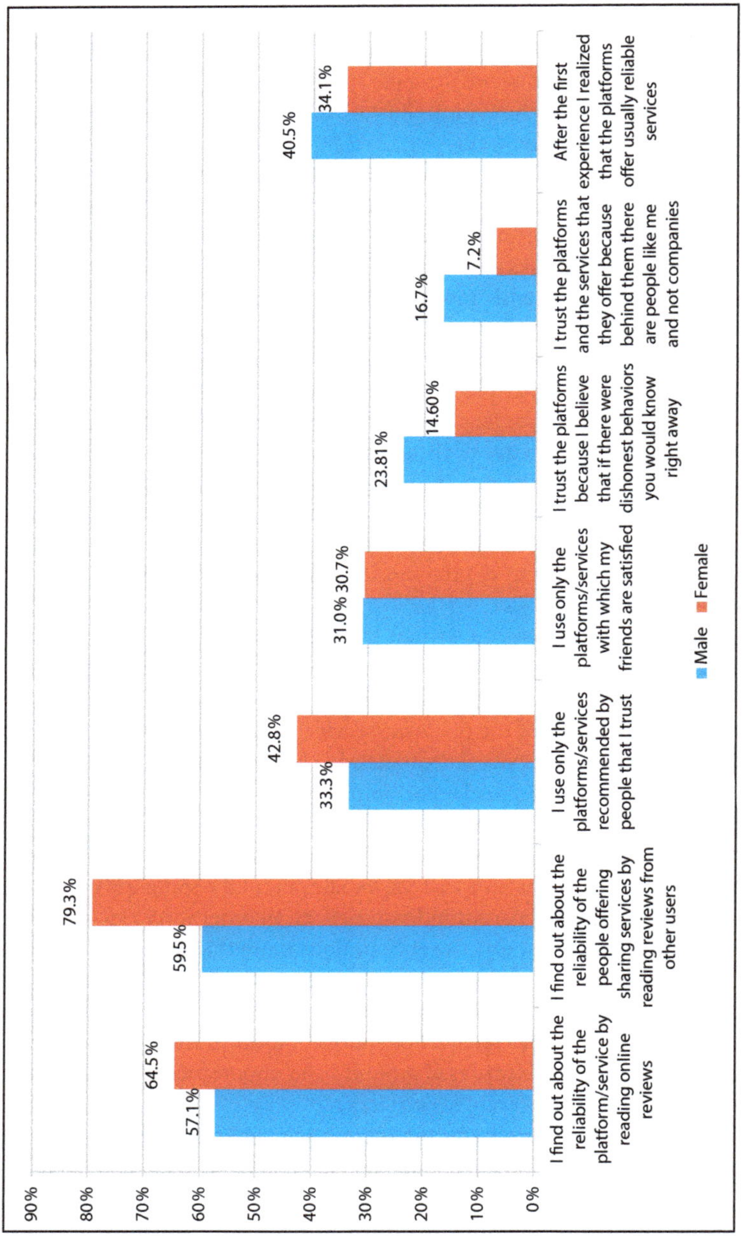

Universe: users of at least one practice. Sum of "entirely agree" and "mostly agree".

Table 8 Variance analysis for items related to platform use (by gender)

D.9 – Sex	I find out about the reliability of the platform/service by reading online reviews	I find out about the reliability of the people offering sharing services by reading reviews from other users	I use only the platforms/services recommended by people that I trust	I use only the platforms/services with which my friends are satisfied	I trust the platforms because I believe that if there were dishonest behaviors you would know right away	I trust the platforms and the services that they offer because behind them there are people like me and not companies	After the first experience, I realized that the platforms offer usually reliable services
Male	1.522	1.533	1.456	1.071	1.102	0.974	1.118
Female	1.044	0.880	1.078	1.075	1.072	0.907	0.669
Tot.	1.158	1.098	1.171	1.069	1.088	0.921	0.776

Universe: users of at least one practice.

Figure 9 Thinking about why you did *not* use these new activities/services, express your level of agreement with each of the following statements

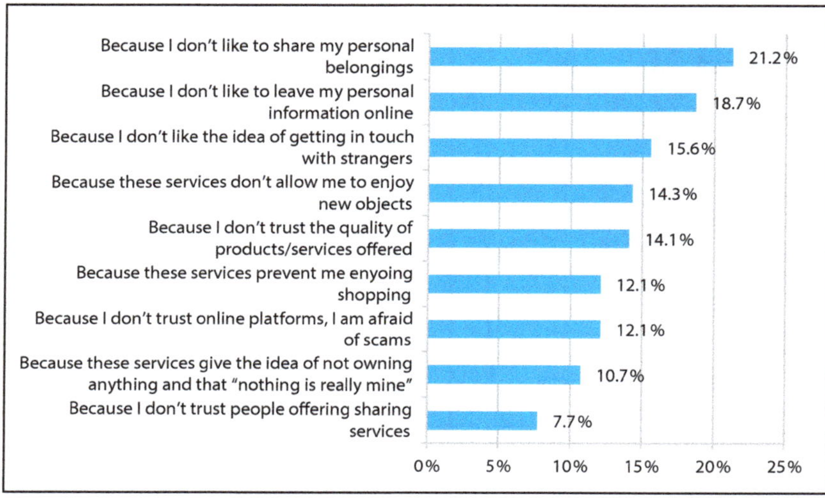

Universe: all respondents. Sum of "entirely agree" and "mostly agree".

Figure 9 shows that the level of agreement (sum of "entirely agree" and "mostly agree") is generally low. 21.2 % of respondents do not like the idea of sharing their personal belongings and 18.7 % do not like to leave their personal information online. 14 % of the sample fear the poor quality of products or services offered, 12 % are afraid of some scam, and just 7.7 % have trust issues with people offering SE services.

The TNS (2015) survey – on 1,000 people representative of the Italian population – only partially confirms these results. Indeed, 15 % of Italians do not use SE services because of trust issues (they fear a scam), 7 % do not like the idea of not owning anything, and 11 % are against the SE concept. Moreover, according to an 2016 Altroconsumo survey[10] (carried out with four other consumers' associations), 19 % of European consumers do not trust people offering SE services, a percentage higher than the 7.7 % in the sample of young people. It can be argued that age differences between the two samples account for the different levels of trust. The European research sample is a representative sample of the European popu-

10 The survey was carried out on a sample of around 9,000 European consumers with a view to assessing the spread of SE services in Europe.

lation, and the core target of SE users is 37 years old. Thus, students seem to trust strangers more than older people do.

If data are stratified by gender of the respondents (Fig. 10), the items showing a sizeable gap in favor of male (with a gap of almost 5 percentage points, summing "entirely agree" and "mostly agree") are those related to the possession of a good, rather than those related to the concept of trust. More specifically, 16.4 % of boys and just 10.7 % of girls agree with the item "Because these services prevent me enjoying shopping". 14.9 % of male students and 9.3 % of female choose the item "Because they are services that give the idea of not owning anything and that 'nothing is really mine'". Finally, almost 18 % of males and 13 % of females agree with "Because these services don't allow me to enjoy new objects". It is plausible to speculate that the male part of the sample is more faithful to a traditional form of consumption behavior, which transforms the purchase – and thus the ownership – of a good into an identity statement. Conversely, the female part of the sample is more concerned with trust issues. Among the trust-related items, the one showing the largest gap between female and male answers is girls' fear of online scams.

Indeed, according to table 9, the item with a lower variance, both for men and women, is the same as the one with which there is a lower level of agreement.

Owyang and Samuel (2015) point out that SE growth may alter the consumer-brand relationship. Indeed, "while it might seem like the move to a peer-to-peer economy would make branding irrelevant, big brands actually matter more than ever. The whole point of a brand is to serve as a promise for product or service quality. In the collaborative economy, customers turn to brands to determine whether a transaction is trustworthy" (p. 17). Following this lead, many brands, like Marriott and BMW, have already implemented new ways of combining the ease and convenience of SE services with the reliability of the best known brands.

The chi square test (Table 10) confirms a significant relationship between gender and "I find out about the reliability of the people offering sharing services by reading reviews from other users". This is the same item which shows the greatest difference between male and female respondents (Table 8).

As far as the items related to the motivations for not using SE practices are concerned, the chi square test (Table 11) shows that the only item having a significant relationship to gender is "Because these services prevent me enjoying shopping". However, among the other non-significant items, "Because I don't trust people offering sharing services" shows the highest chi square value.

Figure 10 Thinking about why you did NOT used these new activities/services, express your level of agreement with each of the following statements (according to gender)

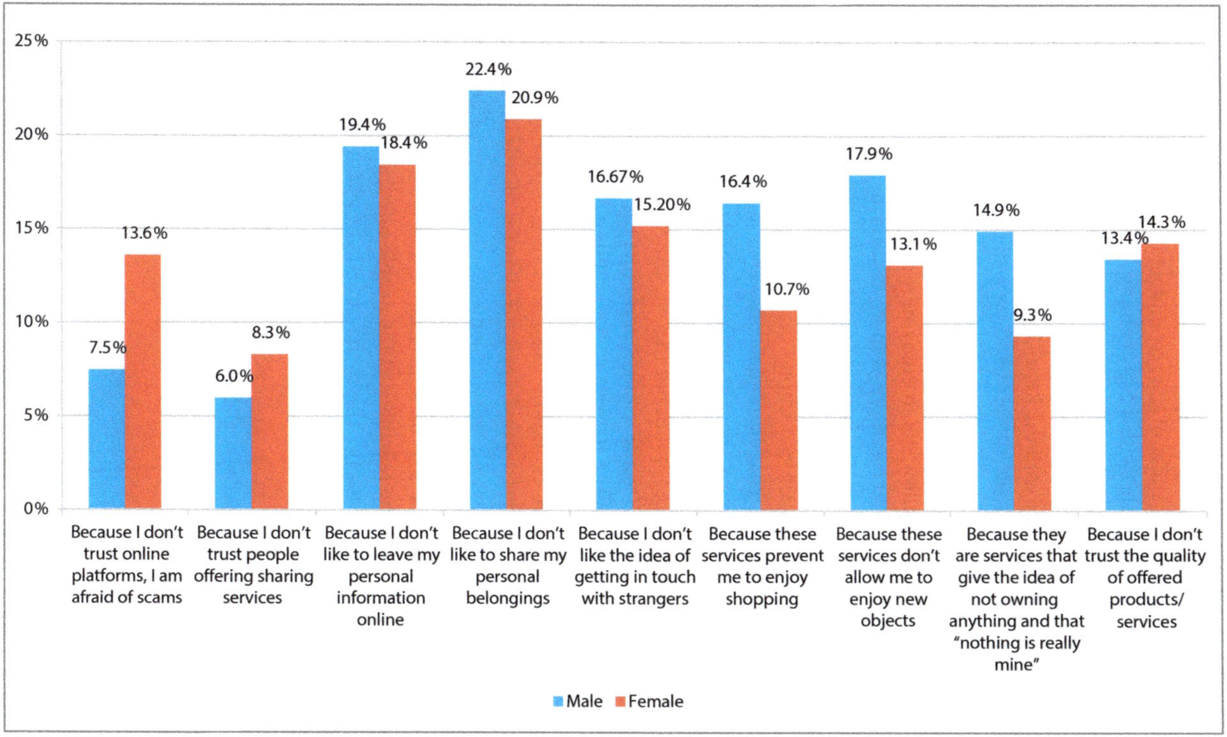

Universe: all respondents. Sum of "entirely agree" and "mostly agree".

Table 9 Variance analysis for items related to reason for not using the platform (according to gender)

D.9 – Sex	Because I don't trust online platforms, I am afraid of scams	Because I don't trust people offering sharing services	Because I don't like to leave my personal information online	Because I don't like to share my personal belongings	Because I don't like the idea of getting in touch with strangers	Because these services prevent me enjoying shopping	Because these services don't allow me to enjoy new objects	Because they are services that give the idea of not owning anything and that "nothing is really mine"	Because I don't trust the quality of products/services offered
Male	0.999	0.862	1.526	1.497	1.717	0.964	1.362	1.087	0.738
Female	1.186	0.861	1.276	1.300	1.320	1.079	1.274	1.054	1.149
Tot.	1.142	0.858	1.333	1.346	1.407	1.050	1.289	1.057	1.061

Universe: all respondents.

Table 10 Chi square test on gender with items related to platform use

	Value	Sign.
I find out about the reliability of the platform/service by reading online reviews	6.997	0.136
I find out about the reliability of the people offering sharing services by reading reviews from other users	14.661	0.005
I use only the platforms/services recommended by people that I trust	5.185	0.269
I use only the platforms/services with which my friends are satisfied	2.968	0.563
I trust the platforms because I believe that if there were dishonest behaviors you would know right away	6.947	0.139
I trust the platforms and the services that they offer because behind them there are people like me and not companies	4.207	0.379
After the first experience, I realized that the platforms offer usually reliable services	5.811	0.214

Universe: users of at least one practice.

Table 11 Chi square test on gender with items related to reason for not using the platform

	Value	Sign.
Because I don't trust online platforms, I am afraid of scams	3.759	0.440
Because I don't trust people offering sharing services	7.566	0.109
Because I don't like to leave my personal information online	3.356	0.500
Because I don't like to share my personal belongings	2.776	0.596
Because I don't like the idea of getting in touch with strangers	2.079	0.721
Because these services prevent me enjoying shopping	11.148	0.025
Because these services don't allow me to enjoy new objects	1.782	0.776
Because they are services that give the idea of not owning anything and that "nothing is really mine"	2	0.736
Because I don't trust the quality of products/services offered	1.905	0.753

Universe: all respondents.

Conclusions

The paradigmatic shift from owning to sharing is a process that started some years ago. As Rifkin (2014) points out, it is a natural and inevitable feature of contemporary society, along with other examples of change. As Levine (2009) stated "sharing is to ownership what the iPod is to the eight-track, what the solar panel is to the coal mine. Sharing is clean, crisp, urbane, postmodern; owning is dull, selfish, timid, backward". According to several studies, the sharing economy and collaborative consumption relies on Internet technologies that allow people to connect and to make better use of goods, skills, and other useful things. Moreover, they allow people to communicate in a peer-to-peer way (Stokes, Clarence, Anderson and Rinne 2014). Along with the lessening of the digital divide, the economic crisis of recent years has fueled the diffusion of SE practices, letting people rethink most of their consumption attitudes and encouraging a more thoughtful and knowledgeable use of goods and services. In this sense, sharing platforms help consumers to transform their values into behaviors.

As far as Italy is concerned, recent studies (Ipsos 2014; Capeci 2015; Eurisko 2016) highlight the fact that the Italian population inclines toward SE practices. The same sources underline a deep gap between the intention to engage in these practices and the actual use of the platforms. Indeed, compared to other European countries, Italy is the country with one of the lowest levels of collaborative forms of consumption (Pais and Mainieri 2015). Likewise, according to Altroconsumo survey (2016), in Italy the penetration of SE services is lower than in other countries (62 % compared to the 72 % of the European mean). Almost all the Italian surveys mentioned point out that young adults between 18 and 34 years old – often labelled as Millennials – are the group most engaged in SE practices. Compared to these findings, the study presented, which focuses on younger individuals, highlights a few significant differences.

University students, aged between 20 and 25 years, claim to know SE practices, but their usage is lower than the national average, which is 12 % (Pais and Mainieri 2015). Significantly, both knowledge and usage of couchsurfing (7.8 %) and ridesharing (9.8 %), the two most famous practices widely discussed by the media system, are lower. As authors have observed, for some practices such as carsharing, adoption rates appear significantly higher among students enrolled in master's courses (mostly born between 1991 and 1993). Thus, age is once again shown to be a relevant factor. As far as the other variables are concerned, gender is not significant in most of the analyses, perhaps because the female part of the sample is overrepresented. Moreover, a discussion based on cultural capital (Bourdieu 2001 [1979]) is of little significance, since all respondents are currently pursuing a university degree in the same academic field (communication). Furthermore, compared to the overall population, students are more sensitive to cost and sustainability issues. They are less motivated by novelty and the possibility of new experiences (Capeci 2015).

The mantra that "the more information, the better" (Finley 2013) is a key to minimizing the risk of online transactions, but the lack of trust is not the main cause of the limited use of SE practices. Indeed, students are keen to verify the reliability of people and platforms. Like the majority of the Italian population (Eurisko 2016), students believe that online reviews foster the collaborative platform's reputation. In the 21st century, reputation is a person's most valuable asset: "new trust networks, and the reputation capital they generate, will reinvent the way we think about wealth, markets, power and personal identity in ways we can't yet even imagine" (Botsman 2012). Most of the collaborative consumption platforms offer trustful mechanisms to evaluate user reputation and minimize the risk of sharing among strangers. Moreover, many sites offer some forms of insurance in order to reduce the risk of any damage or loss. As Wagner et al. (2015) point out, offering insurance protects against the risk of financial losses. It can reduce people's reluctance to engage with a sharing platform and to offer their valuable assets to other peers. Indeed, by posting information and rating users, consumers help each other and increase their sense of belonging to a community. As Rosen et al. (2011) have noted, there is a positive relationship between trust and sense of belonging; to be engaged in a community is one of the main drivers of participation in sharing activities (Möhlmann 2015). Moreover, according to Finley (2013), aesthetic and functional aspects have a large impact on the perceived trustworthiness of the website, as users tend to rely heavily on interface design. A user-friendly site offering high-quality images of products is a guarantee of a professional service (Corritore, Kracher and Wiedenbeck 2003).

Conversely, trust-related items which are among the likely causes for not engaging in SE practices (Fig. 9), show low levels of agreement. On the one hand, this evidence points to the persistence of a traditional relationship with the world of goods. On the other hand, it highlights a lack of knowledge of these platforms. For non-SE users, "I don't like to share my personal belongings" is the most agreed upon item (21.2 %) – they have a limited interest in the whole concept of sharing, rather than trust issues. Moreover, limited knowledge of most of SE practices (Fig. 3) is a good reason for their infrequent use. Thus, if according to Belk (2014a, p. 1599), "it would be folly to ignore sharing and collaborative consumption as alternative ways of consuming and as new business paradigms", the main problem for Italian platforms is reaching a number of users (Botsman and Rogers 2011) large enough to trigger a virtuous circle (Pais and Mainieri 2015) and to compete with conventional shopping practices. Indeed, to succeed, these platforms need a huge volume of transactions able to satisfy peoples' requests and to encourage them to come back. Platforms have to communicate their service and let people understand their meaning. Authors seek more accessible information and an increase in communication so as to deepen the knowledge of the phenomenon and the availability of the platforms. With this purpose in mind, BlaBlaCar has already produced two web series (BlaBlaCar Road Movie) in which commercial partners have inserted their product logos (Boroni 2015).

The media have hotly debated the implications of the sharing economy. The huge success of Airbnb, which without owning a single hotel has become the largest hotel chain in the world, has forced competitors to fight back by modifying their offering but also by changing the law, as recent events in New York attest (Rosenberg 2016). Among other issues, central to the SE debate is the question of whether the SE is bringing more paid opportunities to a greater number of people, or whether its final effect is to shift traditionally secure jobs and to create a landscape of part-time, low-paid work. The "dark side" (Bowman 2016) of this new form of economy is starting to emerge as people ask themselves whether Collaborative Consumption is a way of escaping the power of big corporations, or if, on the contrary, it destroys steady jobs and consumer rights by fostering a growing class of workers in precarious employment (Olmeda 2016). Since the main aim of this research is to explore the perception of SE practices and their usage among young people in Italy, it is feasible to conceive of further steps that might involve different age groups and people of different countries, to enable international comparison. Moreover, further analysis could analyze the relevance of sociodemographic variables in order to refresh the current conception of typical users of SE services.

References

Albinsson, P. A., & Perera, Y. B. (2012). Alternative marketplaces in the 21st century: Building community through sharing events. *Journal of Consumer Behaviour, 11*, 303–315.

Altroconsumo (2016). *Io condivido. Quando il valore è partecipato.* Retrieved from https://www.altroconsumo.it.

Bauman, Z. (2007). *Consuming Life.* Cambridge: Polity Press.

Beck, U. (1992). *Risk Society: Towards a New Modernity.* London: Sage.

Beldad, A., de Jong, M., & Steehouder, M. (2010). How shall I trust the faceless and the intangible? A literature review on the antecedents of online trust. *Computers in Human Behavior,* 26, 857–869.

Belk, R. (1988). Possessions and the Extended Self. *The Journal of Consumer Research,* 15(2), 139–168.

Belk, R. (2010). Sharing. *Journal of Consumer Research,* 36, 715–734.

Belk, R. (2014a). You are what you can access: Sharing and collaborative consumption online. *Journal of Business Research*, 67, 1595–1600.

Belk, R. (2014b). Sharing Versus Pseudo-Sharing in Web 2.0. *Anthropologist*, 18(1), 7–23.

Bella, M. (1988). *La struttura dei consumi privati: un'analisi regionale*. Milano, FrancoAngeli.

Bicchieri, C., Duffy, J., & Tolle, G. (2004). Trust among Strangers. *Philosophy of Science*, 71(3), 286–319.

Böckmann, M. (2013). *The shared economy: It is time to start caring about sharing: value creating factors in the shared economy*. Paper presented at the 1st IBA BT Conference, Enchede, The Netherlands, University of Twente, retrieved from http://thelovettcenter.com/wp-content/uploads/2014/11/bockmann-shared-economy.pdf.

Boroni M. (2015, september 28). La webserie di BlaBlaCar e le nuove forme di pubblicità della sharing economy. *Wired*. Retrieved from http://www.wired.it/author/mboroni/.

Botsman, R. (2012, June). The currency of the new economy is trust. Retrieved from http://www.ted.com/talks/rachel_botsman_the_currency_of_the_new_economy_is_trust.html/.

Botsman, R., & Rogers, R. (2011). *What's Mine Is Yours. The Rise of Collaborative Consumption*. London: HarperCollins.

Bourdieu, P. (2001 [1979]). *La distinzione. Critica sociale del gusto*. Bologna: Il Mulino.

Bowman, R. J. (2016). The Dark Side of 'The Sharing Economy'. September 2. Retrived from http://www.supplychainbrain.com/content/general-scm/business-strategy-alignment/single-article-page/article/the-dark-side-of-the-sharing-economy/

Brown, E., Dury, S., & Holdsworth, M. (2009). Motivations of consumers that use local, organic fruit and vegetable box schemes in Central England and Southern France. *Appetite*, 53(2), 183–188.

Cairns, K., Johnston, J., & MacKendrick, N. (2013). Feeding the 'organic child': Mothering through ethical consumption. *Journal of Consumer Culture*, 13 (2), 97–118.

Campbell Mithun (2012). National study quantifies reality of the "sharing economy" movement. Retrieved from http://www.campbell-mithun.com/678_national-study-quantifies-reality-of-the-sharing-economy-movement.

Capeci, F. (2015). Sharing economy italiana: chi, cosa, quanto… quando e dove? retrieved from http://www.tnsglobal.it/sites/default/files/FCapeci_TNS_%20 Sharing_Economy_Italiana_ebook.pdf.

Censis (2015). *Vita da Millennials: web, new media, start up e molto altro. Nuovi soggetti della ripresa italiana alla prova*, retrieved from file:///F:/16_sharing/Vita_da_Millennials_-_Testo_completo.pdf.

Coley, D., Howard, M., & Winter, M. (2009). Local food, food miles and carbon emissions: A comparison of farm shop and mass distribution approaches. *Food policy*, 34(2), 150–155.

Coomber, R. (1997). Using the Internet for Survey Research. *Sociological Research Online*, 2(2). Retrieved from http://www.socresonline.org.uk/socresonline/2/2/2.html.

Corritore, C.L., Kracher, B., & Wiedenbeck, S. (2003). On-line trust: concepts, evolving themes, a model. *International Journal of Human-Computer Studies,* 58(6), 737–758.

Cowan, M. (2015, April 14). BlaBlaCar has turned ride-sharing into a multi-million-euro business. *Wired,* retrieved from http://www.wired.co.uk/article/blablacar.

David, S., & Pinch, T.J. (2005). Six degrees of reputation: The use and abuse of online review and recommendation systems, retrieved from http://papers.ssrn.com/sol3/papers.cfm?abstract_id=857505.

de Singly F. (1994). Vent'anni di trasformazioni della famiglia e dell'identità femminile. *Studi interdisciplinari sulla famiglia,* 13, 119–138.

Eurisko (2016), *Annual seminar, Disintermediazione tra empowerment individuale e condivisione digitale. Rischi e opportunità della nuova realtà sociale e di consumo,* http://gfk-deliverycenter.it

Finley, K. (2013). *Trust in the Sharing Economy: An Exploratory Study.* Centre for Cultural Policy Studies, University of Warwick, retrieved from https://www.heartland.org/policy-documents/trust-sharing-economy-exploratory-study.

Hamari, J., Sjöklint, M., & Ukkonen, A. (2015). The Sharing Economy: Why People Participate in Collaborative Consumption. *Journal of the Association for Information Science and Technology.* doi: 10.1002/asi.23552.

Handcock, M.S., & Gile, K.J. (2011). Comment: On the concept of snowball sampling. *Sociological Methodology,* 41(1), 367–371.

Handy, S.L., & Clifton, K.J. (2001). Local shopping as a strategy for reducing automobile travel. *Transportation,* 28(4), 317–346.

Harrison, R., Newholm, T., & Shaw, D. (2005) (eds). *The Ethical Consumer.* London: Sage.

Ipsos Public Affairs (2014). *La Sharing Economy in Italia.* Retrieved from http://www.collaboriamo.org/media/2014/07/Sharing_Economy_in_Italia.pdf

Istat (2014). *Il benessere equo e sostenibile in Italia.* Retrieved from http://www.istat.it/it/files/2014/06/Rapporto_Bes_2014.pdf

Lauterbach, D., Truong, H., Shah, T., Adamic, L., & Arbor, A. (2009). Surfing a web of trust: Reputation and Reciprocity on CouchSurfing.com. *International Conference on Computational Science and Engineering,* 346–353. doi: 10.1109/CSE.2009.345.

Levine, M. (2009, March 5). Share My Ride, *The New York Time.* Retrieved from http://www.nytimes.com/2009/03/08/magazine/08Zipcar-t.html?_r=0.

Lubin D.A., & Esty D.C. (2010, May). The Sustainability Imperative. *Harvard Business Review,* retrieved from https://hbr.org/2010/05/the-sustainability-imperative

Malanima, P., & Daniele, V. (2007). Il prodotto delle regioni e il divario Nord-Sud in Italia (1861–2004). *Rivista di politica economica,* 267–315.

Mayzlin, D., Dover, Y., & Chevalier, J. (2014). Promotional reviews: An empirical investigation of online review manipulation. *The American Economic Review,* 104(8), 2421–2455.

Möhlmann, M. (2015). Collaborative consumption: Determinants of satisfaction and the likelihood of using a sharing economy option again. *Journal of Consumer Behavior,* 14, 193–207.

Nelson, M. R., & Rademacher, M. A. (2009). From trash to treasure: Freecycle.org as a case of generalized reciprocity. *Advances in Consumer Research,* 36, 905–906.

Olmeda C. (2016) (eds.). Collaboration or business. Collaborative consumption: From value for users to a society with values. Retrived from: https://www.altroconsumo.it/.

Owyang J., & Samuel A. (2015). *The new rules of collaborative economy*. Retrieved from https://www.visioncritical.com/wp-content/uploads/2015/11/collaborative-economy-report.pdf

Pais I., & Mainieri M. (2015). Il fenomeno della sharing economy in Italia e nel mondo. *Equilibri,* 1, 11–20. doi: 10.1406/79306

Paterson, M. (2006). *Consumption and Everyday Life*. London: Routledge.

Putnam, R. (2000). *Bowling Alone. The Collapse and Revival of American Community.* New York: Simon & Schuster.

Rifkin, J. (2014). *The zero marginal cost society: The internet of things, the collaborative commons, and the eclipse of capitalism*. London: Macmillan.

Rohe, W. M. (2004). Building social capital through community development. *Journal of the American Planning Association,* 70(2), 158–164.

Rosen, D., Lafontaine, P. R., & Hendrickson, B. (2011). CouchSurfing: Belonging and trust in a globally cooperative online social network. *New Media & Society,* 13(6), pp. 981–998.

Rosenberg Z. (2016). Illegal Airbnb listings in NYC will now incur hefty fines. October 21. Retrived fron http://ny.curbed.com/2016/10/21/13361942/airbnb-illegal-short-term-rentals-fines-nyc.

Scarpellini E. (2011). *Material Nation: A consumer's history of modern Italy*. Oxford, Oxford University Press.

Schmitt, B. (1999). Experiential marketing. *Journal of Marketing Management,* 15(1-3), 53–67.

Schor, J. (2014, October). Debating the Sharing Economy. *GreatTransition Initiative,* retrieved from http://www.greattransition.org/publication/debating-thesharing-economy.

Schor, J., & Fitzmaurice, C. J. (2015). Collaborating and Connecting: The emergence of the sharing economy. In L. Reisch & J. Thogersen (eds.), *Handbook on Research on Sustainable Consumption* (410–425). Cheltenham, UK: Edward Elgar.

Sheth, J. N., Sethia, N. K., & Srinivas, S. (2011). Mindful consumption: A customer-centric approach to sustainability. *Journal of the Academy of Marketing Science,* 39, 21–39.

Sills, S. J., & Song, C. (2002). Innovations in survey research an application of web-based surveys. *Social science computer review,* 20(1), 22–30.

Stokes, K., Clarence, E., Anderson, L., & Rinne, A. (2014). *Making Sense of the UK Collaborative Economy*. London: Nesta.

Stolle, D. (2002). Trusting strangers: The Concept of Generalized Trust in Perspective. *Austrian Journal of Political Science,* 31(4), pp. 397–412. doi: http://dx.doi.org/10.15203/ozp.814.vol31iss4.

Stewart, K. J. (2003). Trust Transfer on the World Wide Web. *Organization Science,* retrieved from http://dx.doi.org/10.1287/orsc.14.1.5.12810.

Sundararajan, A. (2013, January 03). From Zipcar to the Sharing Economy. *Harvard Business Review,* retrieved from https://hbr.org/2013/01/from-zipcar-to-the-sharing-eco.

Tanturri M.L. (2011). Diventare genitori: un viaggio tra i modelli di maternità e paternità. In E. Ruspini (Ed.), *Studiare la famiglia che cambia* (pp. 87–120). Roma: Carocci.

Wagner T., Kuhndt M., Lagomarsino J., & Mattar H. (2015). *Listening to Sharing Economy Initiatives. Report on a Global Survey.* Retrieved from http://www.scp-centre.org/wp-content/uploads/2016/05/Listening_to_Sharing_Economy_Initiatives.pdf

Appendix

Table 12 Items list for the question 7 "Thinking about when you are using one of these new activities/services, express your level of agreement with each of the following statements"

I find out about the reliability of the platform/service by reading online reviews
I find out about the reliability of people offering sharing services by reading reviews from other users
I use only the platforms/services recommended by people that I trust
I use only the platforms/services of which my friends are satisfied
I trust the platforms because I believe that if there were dishonest behaviors you would know right away
I trust the platforms and the services that they offer because behind them there are people like me and not companies
After the first experience, I realized that the platforms offer usually reliable services

Table 13 Items list for the question 8 "Thinking about why you did *not* used these new activity/services, express your level of agreement with each of the following statements"

Because I don't trust online platforms, I am afraid of scams
Because I don't trust people offering sharing services
Because I don't like to leave my personal information online
Because I don't like to share my personal belongings
Because I don't like the idea of getting in touch with strangers
Because these services prevent me to enjoy shopping
Because these services don't allow me to enjoy new objects
Because they are services that give the idea of not owning anything and that "nothing is really mine"
Because I don't trust the quality of offered products/services

Collaborative consumption and trust-building processes in the emerging new food economy

Roberta Paltrinieri and Stefano Spillare

Abstract

The chapter introduces and describes the concept of New Food Economy (NFE), which is defined as the result of the increasing collaboration among citizen-consumers and producers, in order to perform more sustainable, healthy and right ways to produce and consume food. This emerging phenomenon is considered as a reflexive response to the increasing food risks in the light of the recent radicalization of the systemic elements of modernity. In the current global era, indeed, people need to increase awareness and "active-trust", instead of "expert systems" and "system-trust". Starting from these considerations authors delineate a framework to analyse the trust-building processes at the base of collaboration and participation within several Italian Civic Food Networks (CFNs). In doing so, they attempt to suggest how changes in these processes may affect the empowerment ability of the networks. The results of this explorative research, indeed, highlight the risk of a progressive "routinization of trust" in the development of the CFNs. This is a phenomenon that goes along with the need of the NFE to emerge from the niche, but it could affect the grass-roots innovation capacity that is usually attributed to these kinds of networks.

Zusammenfassung

In diesem Kapitel wird das Konzept der New Food Economy (NFE) vorgestellt. Das Konzept definiert sich durch zunehmende Zusammenarbeit zwischen Bürgern, Verbrauchern und Erzeugern. Hierdurch sollen nachhaltigere, gesündere und gerechtere Wege zur Erzeugung und zum Verzehr von Lebensmitteln gefunden werden. Dieses Phänomen wird als eine reflexive Antwort auf die zunehmenden Lebensmittelrisiken im Lichte der jüngsten Radikalisierung der systemischen Elemente der Moderne betrachtet. In der gegenwärtigen globalen Ära müssen die Menschen Bewusstsein und

I. Cruz et al. (Hrsg.), *Contemporary Collaborative Consumption*,
Kritische Verbraucherforschung, https://doi.org/10.1007/978-3-658-21346-6_7

„aktives Vertrauen“ schaffen anstatt auf „Expertensysteme“ und „System-Vertrauen“ zurück zu fallen. Ausgehend von diesen Überlegungen wird ein Analyserahmen skizziert, welcher Vertrauensbildungsprozesse auf der Basis von Zusammenarbeit und Partizipation in verschiedenen italienischen Civic Food Networks (CFNs) erörtert und können Veränderungsprozesse die Empowermentfähigkeit der Netzwerke beeinflussen. Die Ergebnisse dieser explorativen Studie zeigen das Risiko einer fortschreitenden „routinization of trust“ im Laufe der Entwicklung von CFNs. Dieses Phänomen geht mit der Notwendigkeit der New Food Economy einher, aus der Nische hervorzutreten ohne die Basis-Innovationskapazität zu beeinträchtigen, welche normalerweise solchen Netzwerken zugeschrieben wird.

Keywords

New Food Economy, trust, Italian Civic Food Networks

Introduction: the role of collaborative consumption in the emerging new food economy

In recent years, new alternative agro-food networks have become increasingly popular. The rapidly growing spread of these networks – mostly community-based and aimed at facilitating the consumption of local and healthier foods – has been regarded by various authors as a “turn to quality” (Goodman 2003) and the beginning of a “New Food Economy” (NFE) (Winter 2003). The latter is mainly arranged in Alternative Food Networks (AFNs), like farmers’ markets or other forms of direct food supply chains, such as the so-called Food Community Networks (FCNs). These are local and community-based networks based on a more active participation of the local population in the food supply chain (Migliore et al. 2014). These networks are usually not very formalized and rely on mutual collaboration among consumers and between consumers and producers, usually supported by strong ethical motivations. Some examples are Community Supported Agriculture (CSA) in the US, *Association pour le Maintien d’une Agriculture Paysanne* (AMAP) in France, Solidarity Purchasing Groups (SPGs) in Italy and, in some cases, also farmer’s markets (Migliore et al. 2014).

All these specific agro-food systems are mainly based on production, distribution and consumption models that are idealistically and pragmatically contrary to the prevalent agro-business model (Forno and Graziano 2014). In fact, they are very often rooted in typical food activism concerns (Counihan and Siniscalchi 2014), such as food sovereignty, food democracy and food citizenship (Counihan and Siniscalchi 2014; Renting et al. 2012). In brief, they are typical expressions of "grassroots (economic) activism" (D'Alisa et al. 2015), often extending their purposes beyond mere food issues (Grasseni et al. 2013; D'Alisa et al. 2015). For these reasons, Renting et al. (2012) suggest defining them as Civic Food Networks (CFNs), stressing the new ways in which citizens are engaging in food issues and practical food supply networks. Surely, "as part of these newly emerging networks, citizen-consumers, in collaboration with 'citizen-producers', actively reshape their relations with different stages of the food system and start revaluing the (social, cultural, environmental) meanings of food beyond mere commodity and object of economic transaction" (Renting et al. 2012).

The concrete practices enforced by these networks also result in so-called Civic Agriculture (Lyson 2004) whose assumptions oppose the capitalist agro-business model and which can be interpreted as the "agricultural" version of the "civil economy" (Bruni and Zamagni 2007), a third sector economy network aimed at the most suitable management of common goods and community-based welfare services. The term 'civic' is used herein mostly to stress the active involvement of members in the (co)management of the commons (Paltrinieri and Spillare 2015). As a matter of fact, these networks often act as true civic participation "workshops" focused on social innovation in production, distribution and consumption of food (Migliore et al. 2014; Forno and Graziano 2014; Renting et al. 2012). Hence, they are interpreted as a place of "real utopia" (Cucco and Fonte 2015) in which alternative ways of resilient local development are tried out and managed through democratic participation and mutual collaboration, with an intrinsic "morphogenetic ability" (Donati 2013) to create the unexpected.

CFNs may thus be summarized as collaborative projects for the commons, featuring horizontal participatory and cooperative practices which characterize both them and the emerging NFE, as a model of "collaborative consumption" (CC) or, more generally, as a model of "collaborative economy" (European Commission 2016). Recently, the term CC has been popularized by Botsman and Rogers (2010) who, stressing the role of Information and Communication Technology (ICT) as an enabling factor, have described CC as a "set of practices and models that, through technology and community, allow individuals and companies to share access to products, services and experiences." (p. 29) However, it is quite clear that sharing and collaboration are not new concepts. Humans have hunted and gathered in packs, farmed in cooperatives and bartered through trade networks for thousands of years (Rinne et al. 2013).

In fact, long before Botsman and Rogers, Felson and Spaeth (1978) had defined acts of CC generally as "those events in which one or more persons consume economic goods or services in the process of engaging in joint activities with one or more others." (p. 614) The concept of CC used here refers particularly to those economic models of disintermediation of the supply chain – not necessarily driven by new technologies – directly and reciprocally managed by groups of consumers (or consumers and producers) who act together on the basis of common values and reciprocal trust. Specifically, the involvement of citizens as consumers in these kinds of CFNs could be seen as an evolution of critical (or political) consumers' agency (Micheletti 2003). In these cases, indeed, consumers do not yet limit their possible choices to the binary alternative "to buy" or "not to buy" (boycott), nor do they limit it to "buy different" (buy-cott) among the goods offered in the mainstream market. Contrary to these limited options in which "consumers act as consumers" (albeit critically), in the CFNs the agency of consumers goes further, joining with other people in an alternative and collaborative new economic venture (Spillare 2016). This kind of agency helps transform critical consumers into critical citizens (Renting at al. 2012; Graziano and Forno 2012; D'Alisa et al. 2015), even if first they need to escape from conventional thinking and take a leap, at least initially, into an unknown land of reciprocal interaction.

For this reason, trust plays a key role in the ability of citizen-consumers to start and support long-term collaborative and collective actions towards a common, though not yet more precisely specified purpose (Neilson and Paxton 2010). Relational capital, intended mainly as the "quality of relationships" (Donati 2011, 2013) rather than the quantity, is highly important in the gradual process of the constitution of the NFE. It is an open process, still widely unknown, in which the aims and methods, as well as mutual trust, need to be constantly tested and negotiated. The current analysis will accordingly focus on the relationships within the Italian Civic Food Networks (CFNs), paying particular attention to some of the main elements of the trust-building process. Specifically, the ability to create and maintain "active trust" (Giddens 1994) will be considered the core of the process of resilience and empowerment of local communities. Considering the diversity of experiences that characterize the emerging phenomenon of the NFE, we will focus our attention primarily on the contextual and organizational features in order to understand how (from an evolutionary perspective) these changes could strengthen, weaken or simply modify the collaborative agency of citizen-consumers and the possibilities for their long-term collaboration.

Matters of trust in the emerging new food economy

In order to function properly, societies need, of course, a basic level of trust to allow for social interaction. Schütz (1970) and Garfinkel (1963) already described the "natural attitude" of people to take for granted the social reality of their lifeworld, namely a basic trust in the common sense of their familiar context. Niklas Luhmann, for his part, distinguished between "trust" and "confidence" (Luhmann

1988), considering the latter as "the expectations taken for granted in our everyday life", while trust "presupposes a situation of risk" (Luhmann 1988, p. 97), that is to say, a propensity of social actors to take action. According to Luhmann, while dangers have always been a feature of human survival, the category of risk is typical of the modern horizon insofar as it represents a sort of "interiorization" of possible dangers (Luhmann 1988). In other words, during the modern process of individualization, people have gained awareness that the extent of risk depends on human actions and the way in which social systems are created. At the same time, people have become conscious of their inability to face situations individually.

In their everyday life, people usually rely on functional "abstract systems" (Giddens 1990) based on a large range of checks, control authorities and experts, impossible for each single person to disentangle. Thus, we usually take for granted the functioning of the social system, its persistence and regularity. Luhmann defines "system trust" (Luhmann 1979) as a kind of taken-for-granted certainty, and sees it as "confidence" rather than trust, a sort of impersonal basis for social life. In contemporary society, indeed, "trust remains vital in interpersonal relations, but participation in functional systems like the economy or politics is no longer a matter of personal relations. It requires confidence, but not trust" (Luhmann 1988, p. 102). For Luhmann, trust is closely tied to the ability of social actors to act and make decisions that are not taken for granted, that is, the ability to assume the risks concerning the consequences of our actions and interactions.

Facing the growth in "systemic risks" (Beck 1992), people's confidence in abstract and expert systems drops, triggering reflexive strategies of action, including the assumption of the risk inherent in new collaborative interactions. Due to the de-legitimization of nation-state policies and the growing "organized irresponsibility" caused by the end of a certain "linear technocracy" (Beck 1996, 2005), the ability to establish trust relations instead of impossible certainties has become of utmost importance. People tend to shift from system trust to "active trust", interpreted as a contemporary kind of trust that needs to be constantly worked upon in the rather unstable context of late modernity (Giddens 1994). In fact, the lack of confidence in institutionalized systems generates a "sub-political" request on the part of civil society (Beck 1992, 1996, 2005) which should be interpreted as a way of re-embedding basic needs and symbolic meanings through the reappropriation of the material basis of society. It should not therefore surprise us that political issues are now also treated in alternative contexts, such as the market and consumption field (Micheletti 2003; Stolle and Micheletti 2013).

Returning specifically to our subject and considering what has just been stated, one of the main assumptions of the current analysis considers CFNs the result of a "glocal reflexivity" process (Fonte 2013; Spillare 2016) in which trust is doubly involved, both at a macro and micro level.

At a macro level, we should keep in mind that food is currently the intersection of several socio-economic and global environmental issues, as well as a crucial factor of the geo-political order of tomorrow (Counihan and Siniscalchi 2014). Moreover, food is both an essential and a consumer good, which has led the global agro-food industry to take unexpected risks and to recurring food scares (Beck 1992). These top-down structural and systemic push factors related to food safety, as well as a lack of trust in the global agro-food system (system trust), encourage citizen-consumers to re-embed the food economy into their local communities, exploring or discovering new experimental ways to produce and supply food (Magdoff and Tokar 2009; Grasseni et al. 2013, Fonte 2013; Spillare 2016). In doing so, they define a new alternative geography of food and create new opportunities for small enterprises excluded from globalization (Migliore et al. 2014). Meanwhile, at a micro level, trust relationships play a role in the ability of citizen-consumers to start up and sustain bottom-up collaborative food networks, able to provide a common base for collective actions.

In current discussions, we focus our attention mainly on investigating this last micro level, with the aim of better defining the concept of active trust, its role in sustaining long-term collaborations and its ability to support social innovation in the CFNs, improving the sustainability and resilience of local communities.

The concept of 'active trust' and its relevance

According to Guido Möllering, active trust is a new concept that "Giddens has introduced more or less in passing and which the trust literature has not really taken on board yet" (Möllering 2006, p. 356). In revisiting neo-institutional theory from a trust-based perspective, Möllering asserts that trust, and active trust especially, "should be seen as a continuous process of reflexive constitution which requires mutual openness and intensive communication" (Möllering 2006, p. 366). He suggests that active trust is rooted in a continuous process of active participation in which "the basis for the relationship and trust needs to be constantly reproduced" (Möllering 2006, p. 369). It seems to be a type of trust that is constantly negotiable and negotiated, different, for instance, from "swift trust" (Meyerson et al. 1996) based on role expectation in temporary groups, and contrary to the confidence rooted in taken-for-granted rules, roles and routines.

Active trust cannot therefore be considered merely as the active attitude of people engaging other people in a collaborative relationship in order to deal with systemic risks. Rather, it refers to an active and persistent commitment that allows people to maintain common values, visions and aims. Taking an active part in discussions, deliberations and decision-making processes, CFN participants are able to continuously produce and reproduce common symbolic references, thus constantly defining and re-defining "the social context in such a way that the trust required for cooperation becomes possible" (Möllering 2006, p. 370). Briefly, active trust could be described as a "trust-in-the-making" (Möllering 2006) rooted in "making together". It is a kind of "first-hand trust" that involves people in participative and collaborative experiences, able to carry out a recurring familiarization with the "unknown", embodied in collaborative innovation.

With reference to our subject of interest, we notice how this definition of active trust seems to be appropriate for describing the interactions within CFNs and their particular organization models. As several case studies have demonstrated, the governance of CFNs is indeed mostly based on internal direct democracy and active participation in decision-making processes, while the management of everyday activities is often informal, and mostly based on volunteers with interchangeable roles. A friendly atmosphere of conviviality is also strongly encouraged (Sage 2003; Forno and Graziano 2014; Grasseni et al. 2013; Cucco and Fonte 2015). As Ostrom (2010) underlined, communication helps increase "the trust that individuals acquire when promises are made to them in a face-to-face setting" (p. 158) and it contributes to the ability of people to act together for a common purpose. As Migliore et al. (2014) summarized, other important elements that Ostrom identified as supporting collective actions are, for instance, the role of "conditional co-operators" as "trust entrepreneurs" (Möllering 2006), the structure of the network and the presence of a clear "boundary role" that allows for reciprocity and mutual monitoring among the various actors in the network. However, some scholars have recently noticed structural changes in the fabric of social cooperation, highlighting several appreciable shifts in their organizational model. Public policies, along with other internal motivations, have led associations to become more professionalized and market-oriented, with a stronger internal structure, more employees and an improved focus on service (Maloney et al. 2007; Salamon 1997).

In Italy, for instance, recent national and European policies have pushed towards a subsidiarity approach in the delivery of public services, and several associations emerged or were transformed, with the acquiescence of public authorities, in order to replace public institutions in supplying public services (Forno and Polizzi 2012). This situation certainly increases the efficiency and effectiveness of organizations, but it can also weaken their traditional ability to create relational and social capital, i. e. improve the quality of relationships and the widespread network of social ties (intra- and inter-organizationally). Backman and Smith (2000), for instance, briefly describe this paradox with the expression "healthy organizations, unhealthy communities". Civic associations have definitely increased their ability to communicate in the mediated public sphere, i. e. their advocacy abilities and their influence, as well as their ability to boost their membership, but this membership is often reduced to a "check book participation" (Verba et al. 1995; Pattie et al. 2004). This means a lack of engaging and active participation, with a consequent transformation of the nature and shape of trust within the organizations. This change in trust is thus exactly what we want to investigate in this chapter, in order to arrange the key factors for a long-term community-based and innovation-oriented collaboration better able to manage the commons.

Trust-building processes typologies in the CFNs

In line with the neo-institutional approach to trust introduced by Möllering (2006) and in order to better describe trust relationships within CFNs, as well as the possible shift in the trust-building process, we have identified at least three different trust-building models. These are: 1) process-based trust, 2) characteristic-based trust and 3) institutional-based trust. Here the focus is on the trust-building process, which may be based on organizational processes (their functioning, predictability and controls), rather than the characteristics of the people involved (e.g. socio-economic features), or even shared norms and rules. As has been described above, it is quite clear that the trust-building process in CFNs should be interpreted as "process-based", in view of its rootedness in "making together", while it could also be understood as "characteristic-based", due to its source in a background of shared values (such as similarity or familiarity), which facilitate the initial interactions. "Institutional-based trust", instead, should be read as the agreement to common rules and norms but which are constantly re-negotiated through active participation and collaboration (active trust).

Furthermore, another criterion for classifying trust-building processes may be the personal attitude or motivations of the trustor. From this point of view, we identify another three analytical categories: 1) "calculus-based trust", 2) "knowledge-based trust" and 3) "identification-based trust" (Möllering 2006). The first category concerns a rational choice related to the "opportunity" for trust; the second is founded on reciprocal acknowledgement and a history of the trust relationship; the third is a full commitment of the trustor based on common aims, values and visions. All three elements may be present singularly or together, but it is clear that active trust within CFNs is more concerned with "knowledge-based trust" and "identification-based trust" (even if from an open perspective), rather than "calculus-based trust".

The last useful concept, retrieved by Möllering (2006), is "isomorphism", understood "as the process of adapting (systems of) action to match and imitate institutional requirements" (Möllering 2006, p. 360). In this field, we may identify different analytical types of isomorphism: 1) "competitive", 2) "coercive", 3) "mimetic" and 4) "normative". Competitive isomorphism concerns a situation in which "actors generally trust each other because this practice has emerged as more successful than not trusting" (Möllering 2006, p. 361). It is rooted in the frame of maximization of rational choice or related to disinterested evolutionary forces. It clearly leaves little room for real agency, as does the "coercive" type.[1] "Mimetic" and "normative" isomorphism may be considered as two different levels of inner adherence to common rules and values, i. e. a matter of conformism rather than belief. We also should note how the terms "institution" and "institutionalization" are not used by Möllering (primarily) with reference to "a third-party guarantor and enforcer – as they would be in rationalist theories of trust – but as systems of rules and meanings that provide common expectations which define the actors as social beings" (Möllering 2006, p. 360). However, by definition, active trust reveals an open process, constantly practiced by the members of the CFNs through active participation in the formulation of rules, meanings, objectives, and so on. Therefore, isomorphism also results in an open process, which we can define as "active isomorphism". Conversely, to the extent isomorphism is determined by factors different from the active participation of members, we have to see it as a kind of "passive isomorphism", in which the trusted processes and the adherence to goals and meanings no longer relate to active trust but rather to other processes among the competitive, mimetic or normative ones.

Trust-building processes in some Italian case studies: a framework for exploratory research

In the first section of this chapter, we considered the NFE as a wide range of innovative community-based approaches to the food economy based on common values and predominantly managed by consumers and between consumers and producers collaborating with each other in the so-called CFNs. The main aim of our analysis is thus to figure out in which way changes within the trust-building process in the development of the CFNs may affect collaborative consumer agency, and how this could influence the emerging NFE. Hence, we have attempted first to highlight the main characteristics of the CFN phenomenon as it has appeared in the literature, stressing their particular ability to be innovative in the management of the commons, and strengthening the resilience of local communities.

In the second section, we stressed the role of trust in collaborative performance within CFNs. Specifically, we have identified a top-down macro-level in which structural and systemic concerns act as push factors for a renewed active trust. Meanwhile, at a micro level, the maintenance of trust is indicated as the basis of the CFNs' grassroots innovation and capacity for resilience. Different trust-building processes may thus determine a different future form of CFNs, and

1 In fact, participation in a CFN is a free and voluntary choice, thus coercive forces could be seen as being merely rooted in the structural changes underlying growing systemic risks.

hence the emerging NFE. For this reason, we have examined theoretical elements and mentioned a series of analytical categories and heuristic tools related to the trust-building process, useful for recognizing how contextual and organizational features of the CFNs may strengthen, weaken or simply modify (from an evolutionary perspective) trust relationships and the possibilities for a long-term internal and external collaboration.

In light of the above, we will explore some Italian CFNs, paying particular attention to the different ways in which trust-building affects the agency of citizen-consumers. Specifically, the ability to create and maintain active trust is considered the core of the CFNs' ability to improve collaborative innovation for a concrete resilient empowerment of local communities. To pursue these aims, we have applied the case studies approach (Yin 2009), selecting five of the main Italian CFNs[2], chosen through an evaluation of their civic and innovative approaches, notoriety and networking capabilities. We have also been careful to select among different types of organizations, i.e. more or less structured and of different sizes. We have considered, for instance, not just single SPGs, but an "umbrella" association that functions as a node for delivering food from selected farmers to a wider network of smaller SPGs or single families. We have also selected cooperatives in which agriculture is co-managed and the products are shared among members.

All these kinds of CFNs are chosen in order to investigate trust-building processes in different kinds of associations, including those that are more formal and market-oriented, highly specialized and with greater internal structuration. The following table (Tab. 1) presents the selected case studies and is supported by a brief description.

As a first step, each case was studied mainly through an analysis of documents available on the web site of the selected associations or provided by the people in charge of the same organizations. These sources were useful for a primary understanding of the declared mission and vision of the organizations, as well as their general operation. Secondly, we provided semi-structured interviews submitted to the respective organization managers. This second step helped us better understand the actual operation of each organization, the internal and external relationship fabric, and the contingent recent changes in trust relations.

2 The analysis of the reported case studies is the result of part of the PRIN project (Research Projects of Relevant National Interest), entitled "Food consumption practices and policies within sustainable lifestyles in response to the crisis" and developed between 2014 and 2016 by the Department of Sociology and Business Law of the University of Bologna (Italy).

Table 1 Selected Cases

L'Isola che c'è	An association representing the solidarity economy network within the Province of Como (Lombardia).
Poco di buono	A co-op that formally represents the GAS network of Rimini (Emilia-Romagna). Its main aim is to deliver or sell the products of its farmers' network directly to a shop.
Pulmino Contadino	An association that aims to "physically" connect, by bus, groups of consumers (e. g. SPGs) with social farms within the Province of Grosseto, Siena and Livorno (Tuscany).
Arvaia	Similar to the American CSA, it is a farmers' co-op of citizen-consumers located in Bologna (Emilia-Romagna) and based on the "food sovereignty" principle, in which members produce and consume their own products.
Campi Aperti	An association for food sovereignty that organizes and arranges some "critical" farmers' markets around the city of Bologna.

Collaboration and trust at work in Italian CFNs

The most popular CFNs in Italy are certainly the already cited SPGs. They are fundamentally a system of collective purchasing set up by a number of citizen-consumers who cooperate in order to buy fresh food and other commonly used goods directly from local sustainable producers, resulting in fair prices for both sides. Thanks to their explicit critical consumption approach, the local dimension of the economic exchanges, and the direct relationship among consumers as well as between producers and consumers, SPGs have played a leading role in the development of local and sustainable agro-food chains in Italy, for instance, supporting organic farmers even when this was not yet popular.

Economic and ethical issues related to food are usually discussed among the members of these networks. They test possible innovative solutions based on collaborative, collective actions suited to the real and contingent needs of local communities. Engagement of economic subjects strongly characterized by social and ethical principles, and social control by citizen-consumers of the production process, are the real treasures of SPGs, which are often part of wider networks (or districts) of the solidarity economy.

L'Isola che c'è

Within the case studies analysed, an example of these broader networks is *L'Isola che c'è*, an "umbrella" association representing the coordination of SPGs of Como Province (in north-western Italy). This non-profit association originated in 2004 and has contributed to the multiplication of the number of SPGs in the territory. Currently, it is one of the most important solidarity economy networks in Italy and one of the biggest businesses in the alternative food economy. *L'Isola che c'è* promotes and organizes a solidarity fair that counts more than 900 exhibitors and about 130 markets and has also established a permanent store called *Corto circuito*, involving almost six thousand people overall.

Thanks to the size of its network, *L'Isola che c'è* takes an active part in the popularization and awareness-raising of critical and aware consumption and is usually very much involved in promotional events, because they "... have learnt that proper input can successfully draw attention to a topic, enhancing it" (Director of *L'Isola che c'è*).

The most popular topics thus include unfair access to food worldwide, a gap in the distribution of wealth, food waste and its resulting impact on the environment. The workshops organized by *L'Isola che c'è* teach people how to reduce household food waste by recycling or improving food storage. The main aims of the association are to provide a wider awareness of food issues and practical skills for the "sustainability of everyday life" (Director of *L'Isola che c'è*), while providing a common logistics centre for the delivery of food within the network.

This last need emerged due to the increasing size of SPGs and their networks, which presented them with growing logistical and organizational difficulties. New SPGs frequently emerge from groups which have become too large and chaotic (this process is called "gemmation"). These groups require sufficient human resources for proper management; otherwise they are likely to quickly wane, thus subtracting a small portion of members from the original SPG. The direct relationship and active trust within these local associations usually provide the necessary support, know-how and skills to the new SPGs in order to help replicate them. However, when the size exceeds a certain critical parameter, without gemmation, logistical and organizational problems demand rationalization of procedures, like a more direct and less participatory decision-making process, bigger concessions to suppliers, less stringent selection criteria and more favourable attitudes to the needs of "non-active members", i. e., those who do not actively participate in the SPG but are limited to more or less regular purchases.

The slow bureaucracy that affects the SPGs ("since each SPG has its own directors, rules, etc.") and the slow decision-making process ("because everyone can take part in the discussion") could be seen as limits to the growth of the network. Due to this, within the most successful SPGs – as in the case of *L'Isola che c'è* – more effective delivery points or permanent stores may be established, operating as common logistical crossroads. Purchasing and distribution procedures thus tend to be more formal, and some volunteers often become permanent employees. Briefly, we are faced with a more market-oriented management able to spread alternative ways to supply local food to a much larger number of critical consumers, usually non-active members of an SPG. Nevertheless, the size of the network and the consequent increased formality of this "umbrella" organization have generated some problems, especially concerning the relationship with its SPGs.

Internal relationships within the SPGs, which emphasize the voluntary work of the "active members", may sometimes provoke distrust for those who do not actively participate in the group. Non-active members are seen as free riders, i. e., "people who want to benefit from the organisation without either effort or commitment" (Director of *L'Isola che c'è*). This aspect highlights a paradox, since one of the main goals of the SPGs should be precisely to spread more sustainable food supply patterns among as many people as possible. This apparent paradox is understandable precisely in terms of the trust-building process, especially in connection with the stretching of the "boundary role" within the network, which affects the agency of members, with the loss of their collaborative ability.

Active trust thus needs to be progressively substituted with different tools, more suitable to managing trust relations between different groups of consumers and farmers. For this reason, *L'isola che c'è* has carried out the broader Italian experiment of "Participatory Guarantee Systems" (PGS) called "C'è campo"[3] – a flexible tool thanks to which, ideally, some active members of the SPGs are directly involved in the guarantee process, in order to emphasize the collective nature of the controls.[4] In this way, the SPGs "would contribute to the consolidation of the reputation of the producers, conveying a sense of belonging within the network and leading to the formalization of the mutual confidence and cooperation" (Director of *L'Isola che c'è*). However, it could be seen as a process of replacement of trust, from first-hand participation (active trust) to the reliance on the guarantee process (process-based trust).

Poco di buono

Similar to the case of *L'Isola che c'è* and *Corto Circuito*, a permanent store was established by an SPG in the case of the RiGAS of Rimini (on the east coast of Italy). RiGAS is one of the largest Italian SPGs, bringing together over 2,200 members who are regularly supplied by over 60 local ethical producers. Thanks to its size, in 2007 RiGAS established a cooperative company with a permanent shop called *Poco di buono*, which has also become a picking point for "traditional" critical consumers. Some members see the addition of a shop as a compromise with princi-

---

3 C'è campo is the only Italian SPG recognized by the IFOAM, the International Foundation of Organic Agriculture Movement.

4 SPG members belong to "visit groups" made up of three people: a consumer (preferably from a different location and chosen among the candidates of the committees), an agronomist (or expert in zootechnics) and another producer. The visit group fills in the "visit manual" with a specific commission which releases the certificates and then assesses them.

ples opposing SPG values. As the same Director of *Poco di buono* admitted, "the most inflexible members consider this step [...] not pertaining to SPGs, [...] giving in to capitalistic concepts, which have always been regarded as the 'embodiment of evil' ... even if we describe ourselves as an ethical company..." (Director of *Poco di buono*).

The shift from common principles to a market-oriented attitude implies an expansion of collaborative access to SPGs by non-active members, but it may be accompanied by a decreased ability to develop shared meanings and cooperation. With a stable shop and a stable management, "SPGs are likely to weaken, the meetings intended for placing an order or the distribution of food are increasingly rare... moreover there is a tendency to delegate" (Director of *Poco di buono*). This is the reason why the same management of *Poco di buono* always pays attention "to stimulating the SPGs" (Director of *Poco di buono*).

Changes in the management model thus necessarily affect the trust-building process. The latter shifts from active to "mediated" trust, which focuses on the relationships between "customers" and the new "distributors" as third party mediators. In fact, the Director of *Poco di buono* has claimed: "... most of the value of our operation lies in the certification attributed to the products" (Director of *Poco di buono*). It thus becomes the guarantor of the ethical quality of food and farmers, and trust becomes "second-hand" trust, mostly regarded as "reputation."[5]

In other words, the cases of *L'Isola che c'è (Corto Circuito)* and *Poco di buono* show that even within the CFNs we are facing a turn towards a market-oriented process, which consequently affects the trust-building process. It is quite clear that, in this situation, the concept of trust loses some of its potential meaning as active trust. Thus "process-based trust" herein means not just active trust through "making together" and active participation, but rather confidence in the repeatability of the process, while "characteristic-based trust" risks being seen, *in extremis,* as a kind of conformism, mostly driven by social differentiation logics and questions of adequacy.

Pulmino contadino

Another significant example of an "umbrella" organization is provided by *Pulmino contadino,* established in 2013. It is a non-profit association that tries to connect several local social farmers in the provinces of Grosseto, Siena and Livorno in Tuscany with the SPGs of the area, while promoting their development. Currently, this network is not particularly large (comprising approximately thirty associat-

---

5 In fact, Wang and Vassileva distinguish between "trust" and "reputation", describing the latter as "a peer's belief in another peer's capabilities, honesty and reliability based on recommendations received from other peers" (Wang and Vassileva 2003, p. 1).

ed farms), yet it provides an interesting case as it tries to promote the multi-functionality of small farmers, often social co-operatives strongly characterized by a social mission, offering them concrete support in a rural area far from the famous Tuscan agro-production zones (i. e. Chianti) and quite "isolated" from the main industrial and touristic flows (more concentrated along the coast or in heritage cities).

Pulmino contadino is thus able to support sustainable local development that takes environmental and social matters into account. Indeed, as the founder of the association explains: "… on the one hand [we promote] compliance with work and environment standards throughout the whole supply chain, while, on the other hand [we encourage] strong social motivation […] which enforces mutual support within the network, especially for those producers with a strong sense of social commitment." (Director of *Pulmino contadino*). The *Pulmino contadino* organization is quite small and not yet formalized but, as in the cases of *L'Isola che c'è* and *Poco di buono,* a limited group of volunteers manages the entire service. They organize orders and deliveries and maintain contact with farmers and SPGs, functioning as mediators. However, the presence and the activity of *Pulmino contadino* seem to contribute to the reinforcement of the social capital of the area. The delivery bus route represents a sort of physical link between the social subjects of the area, assisting in, for instance, "the constitution of new SPGs that enhance a welfare network for disadvantaged people, able to offer more job opportunities and more occasions for autonomy and social integration" (Director of *Pulmino contadino*).

From this point of view *Pulmino contadino* could be seen as what Möllering (2006) suggested calling a "trust entrepreneur", an active subject who is able to promote and reinforce trust across the social fabric of the area. *Pulmino contadino,* indeed, seems to be able to promote the active involvement of people in its area, contributing to the creation of new groups of consumers. Nevertheless, as in the cases of *L'Isola che c'è* and *Poco di buono,* the ability of these third-party associations to promote active trust within single SPGs remains to be demonstrated. At the moment, we are most likely in the presence of "calculus-based trust" and "competitive isomorphism", due to a lack of alternatives (at most, supported by a growing "knowledge-based trust" as a consequence of an enduring mutual relationship).

Arvaia

The case of *Arvaia* represents an example of better integration of enterprise logic and consumer collaboration. *Arvaia* is an agricultural cooperative of consumers established in 2013 in Bologna. Its main feature "is the independent production of food directly by consumers, […] in compliance with the principles of food sovereignty, i.e. the right of people to define their own food systems" (Director of *Arvaia*). There are more than twenty founding members of *Arvaia*. They are mainly activists in local associations (often stemming from the same local SPG network) who actually support the production costs through a solidarity-based system ("richer people can acquire a larger share"). Thus, production is planned and mainly directed to the consumption requirements of the members. Briefly, in *Arvaia*, citizen-consumers are the founders of the farm and the main consumers of the crops (distributed on a weekly basis). Self-consumption and "coincidence of the role of both producers and consumers represents an immediate guarantee of food quality for the members" (Director of *Arvaia*). Moreover, the cooperative organizes training courses for learning cultivation techniques based on social and agro-environmental practices that are explicit alternatives to the capitalist and industrial model[6]. Moreover, common economic ventures involve members of *Arvaia*, "furthering ideological reasons" and reinforcing the original common values and trust relationship.

The engagement of members in the management of the cooperative traces a clear "boundary" of mutual identity among them, without affecting their networking ability. From a multifunctional perspective, beyond its members, *Arvaia* has expanded its own activities to different fields such as ecological communication, education and "agro-fitness" activities, and has created a large network of relationships with many different associations in the local territory. Several levels of participation are thus involved in the activities of *Arvaia*.[7] Due to its organization, its strong capacity to create internal and external links, and thanks to the support of the Municipality (which has freely granted the land), *Arvaia* "represents a concrete and reproducible experiment of an agro-ecological autonomy system" (Director of *Arvaia*). In a logic of shared responsibility, this experiment demonstrates that CCs based on active trust are a real alternative to agribusiness, and are becoming a benchmark for all the CFNs in the city and beyond. In this case, formal organization does not affect active trust among the members; conversely, it helps reinforce internal collaboration.

Campi aperti

The last case study analysed is *Campi aperti*, an association originating in Bologna with the aim of creating a network of local farmers' markets on a common "politi-

---

6 For example, chemical herbicides are banned, weeds are eliminated by hand or integrated into the crop, while edible species are sown.

7 For instance, *Arvaia* has also become a "teaching farm" where children and adults can participate in agricultural activities, learning something new about ecological principles and healthy food.

cal and agricultural path" towards "food sovereignty" and characterized by "strong opposition to neoliberal global market principles" (Director of *Campi aperti*). Unlike SPGs, where the collaborative and participatory actions are primarily directed at consumers, in *Campi aperti* producers have a large part in the organization and management of the markets, often in cooperation with local public administration. However, within organizations such as *Campi Aperti,* direct relationships with consumers play a key role. Its political and agricultural path covers social aspects and the shared values of sustainability and solidarity (the difference between these kinds of farmers' markets and traditional local markets is specifically based on these assumptions). In this case, farmers' markets also represent actual forms of collaborative economy, where consumers, not surprisingly, are considered "co-producers", a term which "highlights the active role played by consumers in supporting producers" (Director of *Campi aperti*). According to our conceptualization of CC, the term co-producer refers to the "co-responsibility in a common development project of a local and more sustainable agro-economy" (Director of *Campi aperti*). This co-responsibility ideally involves the engagement of consumers in shared social and participatory control of product quality (like the SPGs already mentioned).

Unlike the cases already analysed, *Campi aperti* draws on strong political inspiration. Its objectives are mainly "to provide support to very small farmers, those who practice agriculture more as a form of income support than a full-time job" (Director of *Campi aperti*). Indeed, one of the main topics of *Campi aperti's* political activity is national food safety laws, considered too restrictive for household producers[8]. Reciprocal trust among members is thus reinforced by a common commitment to fighting the dominant socio-economic and political order, which is seen to be in favour of the larger global actors, at the expense of smaller, local ones. It is quite clear that these political issues play a bonding role, contributing to a sense of belonging among participants. However, consumer-citizen participation is often reduced to a supporting role, thus their agency is limited to a more traditional "buycott" action.

Conclusions

The current analysis mainly concerns the changes occurring in trust-building processes during the recent development of CFNs, in order to understand how these may have affected collaborative consumer agency, influencing the future form of

8 Recently, thanks to the political mediation of *Campi aperti,* the issue of food safety in household production was considered in Regional Legislation on Solidarity Economy Networks.

the emerging NFE. The study of CC practices within the developing NFE has allowed for an in-depth analysis of the phenomenon, which focused on a set of conceptualizations about trust. Specifically, CC has been regarded as a reflexive response to the radicalization of some elements of modernity, thus a reaction to the emerging need for trust, social integration and sustainability in the current global era. From this perspective, food consumption reveals itself to be a wide field of experimentation. Thanks to the appearance of several predominantly community-based CFNs characterized by mutual collaboration among consumers and between consumers and producers, innovative ways of managing the food supply chain have been established, enhancing local sustainability and the resilience of local communities.

Thus, in the CFNs, collaboration does not just imply the participation of members in food supply chain management, but rather, the sharing of common values and participation in the shared definition of meanings and objectives. However, growing network size (with related management issues) may undermine the relational capital, i.e. the quality of relationships, within the CFNs. Despite the fact that managers of the more formalized organizations (umbrella organizations) have insisted on their efforts to stimulate SPGs, the turn towards a more market-oriented organization seems to affect mutual trust among the members, with a subsequent lack of motivation among the volunteers.

Active members "tend to delegate" due to a centralization of management and a rationalization of roles and routines, with a subsequent increasing role of non-active members. In other words, the more structured CFNs are able to extend their influence beyond the boundaries of the network, towards large numbers of non-active members, i.e. critical consumers who limit their agency to a collective purchase mainly based on the reputation of the organization. This expansion process should be interpreted as a kind of "passive isomorphism", something like a confidence in process predictability, which is based more on a passive conformity with rules and roles derived by the management or by a routinization of past decisions. Trust-building processes thus shift from a first-hand experience, rooted in active participation and trust, to a second-hand experience mainly based on the "reputation" of the organization and its decision-making and guarantee processes.

Analytically, we could say we are facing a phenomenon of "substitution" and "routinization" of the trust and relational capital of CFNs. The lack of participation and the routinization of active trust risks weakening the collaborative innovation of the CFNs, hence their ability to match action to concrete community needs. In all honesty, it is quite difficult to identify the dependent variable with any precision. It is not clear if an increasing market-orientation has weakened active trust and mutual participation or, conversely, a weakening of active trust has determined the need for a more formal and rationalized organization. The variables are surely related, but, as some evidence seems to suggest, this situation may also relate to the organization's ability to maintain clear boundaries between roles. A distinct boundary between active and non-active members is definitely weakened by overly strong market-oriented turns, reducing the agency of the citizen-consumers from a participative role to a mere supporting role.

Furthermore, with regard to the resilience of these communities, it is obvious that a more rationalized and strongly organized association is better able to extend the benefits of an alternative food supply chain beyond the small scale. We have already seen the role of these networks in re-designing a new alternative geography of food, creating new opportunities for small enterprises and with useful results in enhancing collective social capital, especially where the existing local social fabric is frayed. However, in most of the cases analysed, a shift toward a more market-oriented model or other elements that affect active and open involvement seems to feed a collaborative model that is less able to promote the morphogenetic ability of CFNs, that is, their ability to continuously generate innovative and contextual ways of managing the commons. This ability is always related to a participatory open process in which aims, methods and mutual trust need to be constantly tested and (re)negotiated. More studies are necessary to further investigate the role of active trust in the ability of community-based organizations to promote active and collaborative agency among citizen-consumers. Additional research concerning the role of this active attitude in the resilience capability of the communities is also required. We hope this exploratory analysis has contributed to delineating part of the path to be followed.

References

Backman, E., & Smith, S. R. (2000). Healthy Organizations, Unhealthy Communities?. *Nonprofit Management & Leadership*, 10 (4): 355–373.

Beck, U. (1992). *Risk Society: Towards a New Modernity*. London: Sage.

Beck, U. (1996). *The Reinvention of Politics: Rethinking Modernity in the Global Social Order*. Cambridge: Polity Press.

Beck, U. (2001). "Mucca pazza" e la società del rischio globale. *Iride: Filosofia e discussione pubblica*, XIV, 33: 235–241.

Beck, U. (2005). *Power in the Global Age*. Cambridge: Polity Press.

Botsman, R., & Rogers, R. (2010). *What's Mine is Your: the rise of Collaborative Consumption*. New York: Harper Collins Publisher.

Bruni, L., & Zamagni, S. (2007). *Civil Economy. Efficiency, Equity and Public Happiness*. Lang Oxford: Peter.

Counihan, C., & Siniscalchi, V. (2014). *Food Activism: Agency, Democracy, and Economy*. New York: Bloomsbury.

Cucco, I., & Fonte, M. (2015). Local Food as 'real utopia'. A critical perspective. *Conference proceeding 'The Social Meaning of Food'*. Budapest: 16–17 June 2015.

D'Alisa, G., Forno, F., & Maurano, S. (2015). Grassroots (Economic) Activism in Times of Crisis. Mapping the redundancy of collective actions. *PACO*, Issue 8(2): 328–342.

Donati, P. (2011). *Relational Sociology. A New Paradigm for the Social Sciences*. London-NY: Routledge.

Donati, P. (2013). The Added Value of Social Relations. *Italian Journal of Sociology of Education*, 5(1): 19–35.

European Commission (2016). *An European Agenda for Collaborative Economy*. Bruxelles: UE Commission. [Online]: https://goo.gl/K6QDHg

Felson, M., & Spaeth, J. L. (1978). Community Structure and Collaborative Consumption: A Routine Activity Approach. *American Behavioral Scientist*, 21(4): 614–624. [Online]: http://abs.sagepub.com/content/21/4/614.

Fonte, M. (2013). Reflexive Localism: Toward a Theoretical Foundation of an Integrative Food Politics. *International Journal of the Sociology of Agriculture and Food*, (10)3: 397–402.

Forno, F., & Graziano, P. (2014). Sustainable Community Movements Organizations. *Journal of Consumer Culture*, 14 (2): 139–157.

Forno, F., & Polizzi, E. (2012). Tra il mercato e la politica. Modelli organizzativi e strategie d'azione dell'associazionismo in Lombardia. In E. D'Albergo & R. Segatori (eds.). *Governance e partecipazione politica*. Milano: Franco Angeli.

Garfinkel, H. (1963). A conception of, and experiments with, 'trust' as a condition of stable concerted actions". In O. J. Harvey (ed.). *Motivation and Social Interaction* (p. 187–238). New York: The Ronald Press Company.

Giddens, A. (1990). *The Consequences of Modernity*. Stanford: Stanford University Press.

Giddens, A. (1994). Risk, trust, reflexivity. In U. Beck, A. Giddens & S. Lash (eds), *Reflexive Modernization* (p. 184–97). Cambridge (UK): Polity Press.

Goodman, D. (2003). The quality turn and alternative food practices: reflections and agenda. *Journal of Rural Studies*, 19: 1–7.

Grasseni, C., Forno, F., & Signori, S. (2013). Beyond Alternative Food Networks: An Agenda for Comparative Analysis of Italy's Solidarity Purchase Groups (GAS) and Districts of Solidarity Economy (DES) vis-à-vis US Community Economies. Draft paper prepared for the UNRISD Conference '*Potential and Limits of Social and Solidarity Economy*'. 6–8 May 2013. Geneva (Switzerland). [Online]: https://goo.gl/FZdE3E.

Graziano, P. R., & Forno, F. (2012). Political consumerism and new forms of political participation: The gruppi di acquisto solidale in Italy. *The ANNALS of the American Academy of Political and Social Science*, 644(1): 121–133.

Lyson, T. A. (2004). *Civic Agriculture: Reconnecting Farm, Food, and Community*. Hanover, London: University Press of New England.

Luhmann, N. (1979). *Trust and Power*. Chichester: Wiley.

Luhmann, N. (1988). Familiarity, confidence, trust: Problems and alternatives. In D. Gambetta (ed.), *Trust: Making and Breaking Co-operative Relations* (p. 94–107). Oxford: Basil Blackwell.

Magdoff, F., & Tokar, B. (2009). Agriculture and Food in Crisis. An Overview. *Monthly Review*, 61(3): 1–16.

Maloney, W. A., Van Deth, J. W., & Roßteutscher, S. (2008). Civic Orientations: Does Associational Type Matter?. *Political Studies*, 56(06): 261–287.

Meyerson, D., Weick, K .E., & Kramer, R. M. (1996). Swift trust and temporary groups. In R. M. Kramer & T. R. Tyler (eds.), *Trust in Organizations* (p. 166–95). Thousand Oaks (CA): Sage.

Micheletti, M. (2003). *Political Virtue and Shopping: Individuals, Consumerism and Collective Action*. London: Palgrave.

Migliore, G., Forno, F., Guccione, G. D., & Schifani, G. (2014). Food Community Networks as Sustainable Self-Organized Collective Action: a Case Study of a Solidarity Purchasing Group. *New Medit*, 4: 54–62.

Möllering, G. (2006). Trust, institutions, agency: towards a neoinstitutional theory of trust. In R. Bachmann & A. Zaheer (eds.), *Handbook of Trust Research*. Cheltenham (UK), Northamptonn (MA) (USA): Edward Elgar: 355–376.

Neilson, L. A., & Paxton, P. (2010). Social Capital and Political Consumerism: A Multilevel Analysis. *Social Problems*, 57(1): 5–24.

Nicosia, F. M., & Mayer, R. N. (1976). Toward a Sociology of Consumption. *Journal of Consumer*, 3: 65–75.

Ostrom, E. (1990). *Governing the commons. The evolution of institutions for collective action*. Cambridge (UK): Cambridge University Press.

Ostrom, E. (2010). Analyzing collective action. *Agricultural Economics*, 41(s1): 155–166.

Paltrinieri, R., & Spillare, S. (2015). Well-being Shift through Healthy Eating: from Organic Consumption to a Paradigm of Alternative Local Development. *Rivista di studi sulla sostenibilità*, 2: 83–94.

Pattie, C., Seyd, P., & Whiteley, P. (2004). *Citizenship in Britain: Values, Participation and Democracy*. Cambridge: Cambridge University Press.

Renting, H., Schermer, M., & Rossi, A. (2012). Building Food Democracy: Exploring Civic Food Networks and Newly Emerging Forms of Food Citizenship. *Int. Jrnl. of Soc. of Agr. & Food*, 19(3): 289–307.

Sage, C. (2003). Social embeddedness and relations of regard: alternative 'good food' networks in South West Ireland. *Journal of Rural Studies*, 19(1): 47–60.

Salamon, L. M. (1993). The Marketization of Welfare: Changing Nonprofit and For-Profit Roles in the American Welfare State. *Social Service Review*, 67(1): 16–39.

Schütz, A. (1970). *On Phenomenology and Social Relations*. Chicago (IL): University of Chicago Press.

Spillare, S. (2016). Agro-ecological Consumerism between System Response and Glocal Reflexivity: the example of Organic Food Consumption. *Quality – Access to Success* (Supplement), 1(17): 580–584.

Stolle, D., & Micheletti, M. (2013). *Political Consumerism. Global Responsibility in Action*. Cambridge: Cambridge University Press.

Verba S., Lehman Sholozman K., & Brady, H. E. (1995). *Voice and Equality: Civic Voluntarism in American Politics*. Cambridge: Cambridge University Press.

Wang, Y., & Vassileva, J. (2003). Trust and Reputation Model in Peer-to-Peer Networks. *Proc. of IEEE Conference on P2P Computing*. (Sweden): 150–157.

Winter, M. (2003). Embeddedness, the new food economy and defensive localism. *Journal of Rural Studies*, 19: 23–32.

Yin, K. R. (2009). *Case Study Research. Design and Methods*. London, New Dheli: Sage.

Shared use and owning of clothes: borrow, steal or inherit

Ingun Grimstad Klepp and Kirsi Laitala

Abstract

This chapter takes a close look at the different forms of sharing based on empirical material on leisure clothing in Norwegian families. We ask what forms of sharing are practiced, which terms are used, and how consumers draw distinctions between them. We find that the forms are numerous and have an established place in Norwegian clothing culture. Sharing within the household, outside the pecuniary market, appears still to be the most important and also the most understudied form of sharing. The literature about clothing consumption is increasingly about sharing, but limited to the new forms. However, sharing is a common form of human interaction, often misunderstood or overlooked. In order to understand the role of collaborative consumption in contemporary society, it is necessary to study both new and old forms of sharing. For a more systematic mapping of these forms a good place to start might be the study of relationships between access and ownership, and between different temporalities.

Zusammenfassung

Dieses Kapitel unterscheidet verschiedene Formen des „Teilens" von Freizeitkleidung welche auf empirischem Untersuchungen in norwegischen Familien basieren. Das Kapitel untersucht welche Formen des Teilens praktiziert sowie welche Begriffe verwendet werden aber auch wie Verbraucher zwischen ihnen unterscheiden. Die Analyse zeigt, dass es zahlreiche Formen des Teilens gibt, welche einen festen Platz in der norwegischen Kleidungskultur haben. Das Teilen innerhalb des Haushalts, also außerhalb von regulären, monetären Märkten, scheint immer noch die wichtigste jedoch auch die am wenigsten erforschte Form des Teilens zu sein. In der Literatur über Konsum von Kleidung wird das Teilen zunehmend diskutiert, jedoch beschränkt es sich oft auf neue Formen. Teilen ist jedoch eine häufig vorkommende Form menschlicher Interaktion, die

I. Cruz et al. (Hrsg.), *Contemporary Collaborative Consumption*,
Kritische Verbraucherforschung, https://doi.org/10.1007/978-3-658-21346-6_8

oft missverstanden oder übersehen wird. Um die Rolle des kollaborativen Konsums in der heutigen Gesellschaft zu verstehen, ist es notwendig, sowohl neue als auch alte Formen des Teilens zu untersuchen. Eine systematische Abbildung dieser Formen des Teilens ist ein adäquater Ausgangspunkt für das Studium von Beziehungen zwischen Zugang und Eigentum und zwischen verschiedenen Temporalitäten des Teilens.

Keywords

clothing, collaborative consumption, sharing

Introduction

Russell Belk opened a discussion of sharing some years ago by quoting Price (1975), stating that "Sharing is the most universal form of human economic behavior" (Belk 2010, p. 715). Belk argues that sharing is central and important, but often overlooked. He suggests that further research is needed on "how contemporary consumers draw distinctions between sharing, lending/borrowing, gift giving and commodity exchange" (Belk 2010, p. 729). Additional researchers have pointed to the need for a more comprehensive framework for understanding how consumption resources circulate, even though they might not share Belk's understanding of the concept of sharing (Arnould and Rose 2016).

Wahlen and Laamanen (2017) argue against what they describe as an "a-historical perspective" in the contemporary debate on collaboration and the economy, where claims of newness are made without referring to empirical studies on change. We agree with this critique and claim that in order to understand the role of collaborative consumption and a sharing ethos in contemporary society, it is necessary to study both new and old forms of sharing, and to what extent they are growing or receding. We aim to contribute to this discussion through a closer look at the different forms sharing takes in clothing consumption, and by discussing the gap between those sharing forms we find and the focus in the literature on collaborative consumption of clothing. We ask what forms of sharing of clothes are practiced, which terms are used, and how consumers draw distinctions between them. Our empirical material is about leisure clothing in families with children. These are, in other words, people who live in households with other people.

Collaborative consumption can be considered "a form of economic collaboration where idling resources are made available – *shared* – with others in a larger community – *the sharing economy*" (Wahlen and Laamanen 2017, p. 1). Our interest, like that of many others, is based on the potential for reducing the environmental impact of clothing consumption. The number of uses per garment is crucial for the overall environmental impact (Fletcher 2008; Laitala, Boks and Klepp 2015), and sharing is one potential way of increasing the number of uses. Thus, we want to know both what motivates the consumer to share and which sharing forms contribute to the clothes being used by more than one person. This will be discussed briefly in a paragraph on "Motivation and consequences" before the conclusion.

In all social settings, clothes are important for how we are perceived (Kaiser 1997). In the same way, the clothes we wear also affect how we feel (Entwistle 2000). This intimate relationship between clothes, the body and the person is a central topic in clothing research. It is therefore not surprising that clothing is perceived as personal property, when this is also the dominant form of understanding of ownership in our culture (Belk 2010). Both the expressions "coming out of the closet" and "being in someone else's shoes" indicate that using others' clothes implies more than just the practical dimensions. Perhaps this is why we do not so easily "see" the many forms of sharing that surround us. Before we present our empirical material, we will look at what earlier research has taught us about which distinctions are relevant for systematizing the sharing of clothing.

Sharing

Polanyi (1957) has identified three major forms of exchange: market exchange, redistribution and reciprocity. Market exchange is direct and often pecuniary, while redistribution is a system of economic exchange with centralized collection and redistribution of goods. Reciprocity refers to the non-market exchange of goods, labour, services or other commodities, either directly with immediate exchange, or by different forms of gift exchange where a return can be delayed (Parry 1986). Characteristics of a gift are that the one offers thanks upon receiving it (Visser 2008) and that the idea of reciprocity is implicit (Mauss 1970). Sharing as an alternative mode of transfer has been tightly linked to the notion of gift-giving, in contrast to the dominant logic of commercial exchange (Wahlen and Laamanen 2017; Widlok 2013). Belk (2010) separates sharing, gift giving, and commodity exchange, but for us the distinction between sharing within or outside commercial exchange will be important.

Sharing in kinship relations is not new (Wahlen and Laamanen 2017, p. 10). Yates discusses sharing in households and says that larger households appear to share resources more effectively (Yates 2016, p. 2). An important distinction for the discussion on sharing of clothes is thus sharing within as opposed to outside the household. The discussion thus becomes a question of how collaborative consumption events "are events located in space", as Felson and Spaeth (1978) put it.

The spatial aspect is linked to the question of "new" and old forms of sharing. A common denominator of collaborative consumption and collaborative economy initiatives is the mediating role of new digital technologies connecting various actors and modes of transfer (Wahlen and Laamanen 2017, p. 1). This has opened up new types of consumer activism and alternative lifestyles, but also new forms of business activity. Sharing mediated through digital technologies makes it easier to share outside the household. In the distinction between sharing within and outside the household, we will therefore look at sharing where new sharing platforms are involved.

In his article about sharing within households, Yates understands sharing as "the way in which people consume resources in common" (Yates 2016, p. 2). A similar simple definition is proposed by Price (1975): "common use or distribution". In our understanding, consumption includes acquisition, use and disposal of goods and services. The relationship between acquisition and use is interesting, because the former goes to who owns the clothing, while the latter touches on who uses them. This is discussed by Belk as "shared access", and Yates states that the "sharing economy" proposes that access rather than ownership is critical, allowing sharing to mean turn-taking, successively shared use, irrespective of whether access costs money (Yates 2016, p. 7). This is a point where there is disagreement on definitions, because "sharing as alternative mode of transfer has been tightly linked to the notion of gift-giving, opposing the dominant logic of commercial exchange" (Wahlen and Laamanen 2017, p. 5; Widlok 2013). However, for our purposes, the significance is in the distinction between access and ownership, as this opens up the discussion of sharing as either shared use or shared ownership or "co-ownership" (Wahlen and Laamanen 2017).

Yates discusses three modes of sharing in domestic economies: successive sharing, simultaneous sharing and shared/divided work (Yates 2016, p. 2). Divided work is relevant for laundry. 83 % of shared UK households say that "all laundry is done together" (Yates 2016, p. 11). This means that parts of the "use" category of clothing consumption, namely laundry, is included in divided work, which will affect other aspects of use of clothing, such as where they are, who has knowledge/control over them and so on (Chapman 1980; Klepp 2006). His successive sharing category is also relevant for clothing. For while a home's interior textiles are shared simultaneously, the clothes are worn on the body, and sharing needs to be done successively. This opens up a discussion on the "timing" of sharing. According to Felson and Spaeth (1978), "timing is an especially important concept for understanding collaborative consumption, since cooperative routine activities require temporal coordination extending beyond the rhythms and tempos of any one individual" (pp. 615–616).

The distinctions we end up with are:

1) Inside and outside commercial exchange
2) Inside and outside the household, and for the latter, use or not of digital sharing platforms
3) Shared use, or shared ownership
4) The sharing's "timing".

We will use these distinctions in the following sections, to look first at earlier research and secondly at what forms of sharing are present in empirical material on leisure apparel in Norway.

Earlier research on sharing of clothes

The amount of clothing consumed in the Western world has increased rapidly. This is a problem for sustainability, and one of the ways to enable change and reduce the amounts is to share more. This potential has increased the interest in sharing within academia and government in recent years. We will first look into what we find on the sharing of clothing inside and outside markets, as well as inside and outside the household, before we look at discussions centering on access, ownership and timing.

Inside and outside the market and the household

Several mappings of material flows within collection and resale of used clothing have been done in connection with political initiatives on sustainability in textiles (Fisher, James and Maddox 2011; Laitala et al. 2012; Morley, Bartlett and McGill 2009; Palm et al. 2014), and they are included in various action plans and roadmaps (Nordic Council of Ministers 2015; WRAP 2015). A survey of ongoing sustainable textile initiatives in the Nordic countries showed that specifically these types of initiatives dominated (Klepp et al. 2015). There is a deficiency in the way reuse is discussed and measured. Most of the above-mentioned reports focus only on the organised forms of clothing collection and resale adopted by businesses and organisations. Research on sharing has pointed out that most of the literature on the area perceives sharing as exchange (Fiske 1991; Price 1975; Woodburn 1998), and when it comes to clothing, the exchange that is focused on is that which takes place in pecuniary markets.

The formal and pecuniary markets, which Polanyi classifies as market exchange and Belk refers to as commodity exchange, include second-hand and vintage stores, used markets, charity thrift stores, and flea markets. Sometimes this is in combination with online versions of these, such as Oxfam and Fretex (Norwegian Salvation Army), which use online sales channels mainly for the very best and most expensive pieces of clothing. Flea markets in Norway are often organized to collect money for organizations and associations, such as school bands (Briel 1999) and therefore approximate to Polanyi's redistribution category. Common to all of the above-mentioned channels is the fact that the clothing changes ownership through a third party. However, Polanyi's market exchange and Belk's commodity exchange are not limited to channels involving a third party. Private direct selling occurs as well, and there are several methods for finding potential new users, such as classifieds on the Internet or in newspapers. Other methods include organizing backyard or garage sales, selling by commission through second-hand stores, renting a stand at flea markets, or selling directly to acquaintances. As opposed to the limited number of studies on informal non-pecuniary clothing circulation, many studies have focused on the numbers of people who have bought second-hand clothing (Ekström, Gustafsson, Hjelmgren and Salomonson 2012; Hiller Connell 2009; Stephens 1985). We find a substantial amount of research with a focus on consumers who acquire vintage or retro clothing (i. e. Brace-Govan and Binay 2010; Cassidy and Bennett 2012; DeLong, Heinemann and Reiley 2005; Jenss 2005).

Gifts, in Belk's terms, and reciprocity or non-market exchange in the informal reuse sector, include a variety of different forms. People hand down items to friends, family, or other acquaintances, participate in clothes swapping parties, inherit clothing, give away things online, use new social media, and so on (Clarke 2000; Cruz-Cárdenas, González and Gascó 2016; Laitala and Klepp 2017). One of the few studies that includes gifts is from the UK (Gracey and Moon 2012). In historical studies of reuse, activities within the household are included and discussed (Ulväng 2012, p. 113). However, in literature on the current situation, we see a tendency for consumption and consumers increasingly to be used, as in the economic literature, to replace the term "customer" (Lien 2004). The consequence is that the studies focus on acquisition and exchange of ownership, and exclude to a large degree other aspects of consumption, such as use.

Owner, user and timing

By use or access we mean the actual wearing of clothes. This is independent of who owns them. There is not much literature on this topic, but changes in ownership in terms of the used market or gift giving are better documented. Inherited clothing shifts owners without monetary payment, like gifts. Whether they are thus "gifts" is negotiable. What is certain is that the clothes get two sequential owners who may also be users.

One study that we have found distinguishes between the different procurement forms based on whether the goods are bought or not, and whether the goods are new or used (1996). These authors have developed a model for the "inventory ownership cycle" which indicates the different consumption stages, from pre-acquisition to disposal, in the same manner as usually specified for the production stages. Although they take into account both market-based forms of change of ownership, and change of ownership beyond the pecuniary market, the model does not take into account the relationship between owners and users. However, Corrigan (1989, 2008), inspired by gift literature, includes it in his study on family clothing consumption. His is the only work we have found that raises questions about ownership and clothing. He was initially more concerned with the different ways of acquisition rather than ownership itself, but observed that between one third and one fourth of the clothes in the families he studied were not purchased by their users. This made him recognize that family members did not have totally individual wardrobes, where all the clothes belonged to a particular person, but that they circulated in different ways. He found the following forms of circulation (Corrigan 1989, p. 524):

A – Gift purchased on market (new item)
B – Gift family made
C – Commodity family made (family member paid to make it)
D – Cast-offs (items no longer worn by the previous owner)
E – Borrowing (with permission)
F – Stealing (borrowing without permission)
G – Self purchased

Corrigan was not concerned with the relationship between access and "ownership", nor with a more fundamental discussion of possible forms of sharing. He was interested, however, in the forms of circulation he found, and discusses them in relation to gender and family relationships.

Most literature we have found is mainly concerned with the change in ownership, and not ownership itself. The form of changes in ownership that is best documented is related to buying used clothes, where topics such as the extent of the phenomenon, motivations and types of consumers are studied (Laitala 2014). This form of change in ownership is not included in Corrigan's list because it did not occur in the families he studied.

Timing is not discussed in the articles we have read. But that does not mean that it is unimportant for the circulation of clothing. Sharing may occur in the form of multiple users using the garment interchangeably, or sequentially by one person using the garment first, followed by a second user. An example is two siblings who share a garment, as opposed to the one inheriting it from the other. This may be referred to as co-ownership. There will be practical and emotional differences between the two forms (Tinson and Nuttall 2007).

Material: six families in Norway

The study of leisure clothing was one of several cases in a research project explaining the rapid growth in leisure consumption (Aall, Klepp, Støa, Engeset and Skuland 2011). We also refer to recent previously published quantitative studies within the area in order to relate the results to a representative selection of Norwegians (Klepp and Laitala 2016; Laitala and Klepp 2014).

The qualitative material consists of wardrobe studies in six families, conducted by Silje Skuland from SIFO. Informants were asked to display the clothes they have for some specific leisure activities: hiking, running, biking and skiing. Interviews focused on each garment, which was also numbered and photographed. The advantage is that the conversation directly links to the clothes, practices and materiality (Klepp and Bjerck 2014). Our desire to get an overview of what the informants had proved, however, to be difficult. First, because there is a fine line between what is used for these activities and for other purposes. A second problem was the storage of clothing. The clothes were located around the daily wardrobes, but also in basements, sheds, attics and cabins.

The families were selected to represent diversity. All families include children, but they have very different economic circumstances and varied interest in the selected activities. All live near Oslo, the capital of Norway, within a short distance of the forest and excellent opportunities for outdoor activities. The eastern parts of the city have a higher percentage of immigrants and low-income families than the west of the city. We will introduce our informant families with pseudonyms. The families that own the least leisure clothing are presented first.

Dina lives alone with her two children in one of Oslo's eastern districts. For her, the good life is not about owning many things. Dina has no car, and her bike has been stolen. The winter activity she is most excited about is sledding. Toboggan and sledges are also used to transport children to kindergarten and school. The family does not own much clothing for specific recreational purposes. They use what they have, or borrow when needed.

Anne, her husband and their two children also live east of Oslo. Anne goes hiking in the woods where they live, and skiing and sledding at their parents' cottage. Their two boys do karate, play soccer and swim, and have clothes for these activities which are not used for anything else. There is some flexibility in the use areas of clothing, for example an everyday winter jacket became a winter hiking jacket after it was well worn. Anne wanted a more physically active life, and a new hiking jacket was a step in that direction.

Elin is alone with two children and lives west of Oslo. Elin goes out in the woods often, both with and without skis. She refers to herself as "a mountain goat". In addition, she does horse-back riding, biking, aerobics and down-hill skiing. She has her own clothing for all activities, but the kids have one set of equipment for winter activities, as well as thin and thick sports coats used for multiple purposes. Her equipment for down-hill skiing is also used for other activities, and ski suits are used on forest trips with campfires, as the many small, burned holes testify. Elin laughs at this and comments that maybe it is time to replace the approximately six year-old jacket.

Beatrice lives on the east side along with her husband and two sons. They are all very active, and clothing for outdoor life has high priority in the family. Although part of it is inherited, the clothing is in good condition and kept in order.

Christian lives with his wife Catrine and son Christoffer west of the city. Father and son are very active with a variety of outdoor activities such as alpine and carving skiing, snowboarding, bandy, ice hockey, football, and downhill biking. Father and son have an excess of new and expensive equipment for all the activities they are engaged in, or have been engaged in earlier. Christian is interested in equipment and knows a lot about brands, styles, security and functionality. For him, the equipment is an important part of the activities.

Lise lives west of Oslo by the edge of the forest with her husband and two children. They want to live a life where outdoor recreation and exercise are central. The husband does football, down-hill skiing, cross-country skiing, cycling and running. Lise's hobbies include down-hill skiing, cross-country skiing, cycling, and aerobics. The children play football and tennis, and do cross-country and down-hill skiing. The family owns a lot of good quality equipment for these activities. The clothes are appreciated in terms of style and appearance, as well as for the joy and good functional properties the equipment provides.

Findings

In the following section, we present the forms of sharing of clothes that our informant families practiced, which terms they used and the distinctions between the forms. We will also discuss these findings in connection with our previous work on Norwegian clothing consumption, and therefore sharing of all types of clothing, not only leisure wear.

Inside and outside commercial exchange

Sharing of clothes within the market may take the form of buying used clothing or leasing/renting, i.e. a form of sharing with or without change of ownership. As in the families Corrigan studied, sharing outside a commercial exchange is the most dominant form in our material.

Buying and selling used clothing is not a major topic in the interviews. When Lise says that they do a cleanup once a year and give what they do not use to charity, her husband adds: "We can sell it on the ski market". She then responds, "Yes, that is also possible", but there is nothing to suggest that they have done so. Christian mentions some safety gear that was sold used. He talks about the low price of new equipment as a reason for not selling more of the used items. The interviewees' limited involvement in selling and buying of used goods is in line with the statistical evidence.

Purchase of used clothing is less common than receiving it. 30 % of the Norwegian respondents said they had received used clothes during the last two years, which is almost twice as many as those who said that they had purchased used clothing (17 %) (Klepp and Laitala 2016). Gifting may be even more common in collectivist societies, such as Ecuador (Cruz-Cárdenas 2013). This is in clear contrast to the attention that the pecuniary market-based solution gets (Fisher et al. 2011).

Rental is not something our informants speak much about, even though renting equipment for alpine activities is an especially well-established practice. Dina sees this as a solution if her son were to be invited on an alpine skiing trip, "then they can rent equipment on the slopes". Dina is also the one who talks about the possibility of lending locally. She uses this solution to argue that it makes it possible for her son to take part in activities even if they do not have equipment, although they have not used this option.

In contrast to this, sharing outside the market sphere is a flourishing activity for our informants, and it is clear that many non-commercial sharing options exist. We found all three categories that Corrigan (1989) found: cast-offs (items no longer worn by the previous owner), borrowing (with permission) and stealing (borrowing without permission).

Borrowing is completely normal in all six families, but Dina borrows the most. Borrowing increases flexibility for various use situations. Dina's large windstopper jacket also acts as a reserve for her son. Her son is not as enthusiastic about this solution. "So you must force him. And he gets so upset and angry too. But it's like if you're going into the backyard and build snowmen, I mean, fold up [the sleeves] and put it on – out!" (Dina).

In addition, those who own the most, borrow. Lise said that she used her husband's clothes when she was pregnant. She would not consider doing it otherwise, not only because of the size, but "we do not use the same colors. I'd never bought similar clothes to his." But for the short period she was pregnant, borrowing was a convenient way to obtain clothing that fitted her.

Borrowing sportswear is a straightforward solution if the activity, and thus also the clothing for it, has a low priority. Anne says that before she bought her own hiking jacket. "I just used the miscellaneous jackets that I had, or borrowed from my husband. I have not had a specialized hiking jacket. I have not really been a hiking person." Borrowing can solve the practical problem of having a jacket to wear. However, owning a hiking jacket contributes to the change she wants: to become a "hiking person".

The difference between borrowing and inheritance may be unclear. The following conversation starts with a question about inheritance, and the father responds by talking about a jacket his son used "when he used to be an active skier. I'd sometimes use it. Sometimes he'd use it as well; so that we interchange certain things. He grows out of things so fast, so I inherit his t-shirts and sweaters". As long as both use the item, the father is borrowing, but when the son no longer uses the garment, it becomes the father's possession and he describes it as inheritance.

Quantitative material on the topic shows that when asked "Do you share or borrow any clothing, shoes or sports equipment with your family or circle of friends?" about 54 % of women report that they shared some, while only 36 % of men report the same (Klepp and Laitala, 2016). Women share most with their siblings, partners and children, while men share mainly with their partners.

There are also major differences between the age groups: only 32 % of adults over the age of 60 share some of these products, while 62 % of young adults (age 18–29) share some. Respondents over the age of 45 share most with their children, 30–44 year-olds mainly with their partners, and young respondents share most with their parents and siblings, although sharing with friends is also common in this age group.

"Stealing" is something in between stealing and borrowing. In Christian's home there is so much equipment that he is not able to keep track of who owns what. Seated in front of a heap of gloves, all of them for cycling, the father says "Some of these are mine and some are Christoffer's. Me, I can't remember." Mother Catrine adds: "Christoffer remembers everything. He [the father] is not allowed to use his [Christoffer's]. It's really strict, 'Have you taken my gloves!'" (Catrine). The father's use of his son's gloves is thus what Corrigan calls stealing, and what our informants call "stealing". We use stealing in quotation-marks to signify borrowing without permission, but with the intention to return after use in line with our informants. Corrigan wrote a lot about this, but found it only between sisters (1989). In his material, none of the fathers "stole". Tinson and Nuttal (2007) called this "covert borrowing" and also found that this was common practice between sisters.

It would be interesting to investigate this further with regard to who "steals" from whom, and how this is perceived. That "stealing" can contribute to better utilization of the clothes is obvious, but does it also contribute to maintaining social relationships? Or the opposite?

Corrigan (1989) uses the term cast-off, and Clarke (2000) calls it hand-medowns. In Norwegian, our informants say "arv", which is probably better translated as *inheritance.* This is a common theme in all interviews. Children inherit from each other and from other older relatives, while parents also inherit from their growing children.

Inheritance is very normal and takes place both within the household and, as we will return to later, outside the household. Inheritance is not just a strategy to exploit out-grown clothing, but also a planned strategy to utilize the resources in the best possible way. Several informants say that they save the children's ski suits just for skiing, so that the kids are not allowed to use them for other activities. "So it will not be used for anything else. It is in quite nice shape. And his sister will inherit it too. It is not too 'boyish'." (Anne).

In connection with discussions of ski suits, it also becomes clear that it is easy to inherit this type of clothing across gender, as long as clothes are fairly gender neutral and "not too boyish", as Anne expressed. This points to an important condition for maintaining the inheritance of children's clothing: that the clothes are not too closely tied to the person's gender or other characteristics.

Inside and outside the household

In our families there is more sharing within than outside the household. Furthermore, there is more sharing with people who are part of the extended family and close friends than with others outside the household. The sharing takes on several forms, like inheritance and borrowing, as we have already seen. However, we would like to discuss another type of sharing which we have not seen previously discussed: stealing, without quotation marks.

Stealing is prohibited, and not something we can expect the informants to tell us that they do. However, being stolen from is a theme in the interviews. The fear of theft affects how kids are allowed to use the clothes. "We've been a bit unfortunate; various jackets and stuff have been stolen from the kids, and it's a bit annoying, so they are not allowed to use everything for school. And if they do and it is expensive, they have to take it with them into the classroom." (Beatrice). The informants who give high priority to outdoor wear make sure that this clothing is not used in school.

Bordering on stealing or finding, Dina describes "container hunting" as looking through other people's waste before it is sent to the recycling center. Like many of the other non-market based categories, stealing is not commonly thought of, but includes opportunities for acquiring things outside the market, although occasionally outside, or at least on the fringes of, the law.

Inheritance is usually understood as something that happens within the family, but when it comes to clothing, it is also common to inherit from others, even though most comes from close relatives. A consumer survey in Norway confirmed what we found in our informant families; that it is most common to inherit from family, followed by inheriting from friends (Laitala and Klepp 2014). However, it is also possible to say that one "inherits" from people you are not related to. Anne has a pair of hiking boots that she describes as "older than me". She explains that "My mum inherited them from a colleague at work… from Mrs. Berg or something like that." This example shows that the shoes are inherited both from close family members and from colleagues. The shoes created a bond with a person that Anne would not otherwise have any relationship with, "Mrs. Berg". This is an aspect of inheritance that it would be interesting to study further.

Another important finding is that women inherit more than men. This is in line with earlier research on sharing that emphasizes this as something belonging to the home and family, and thus concerns women more than men and basically goes on beyond the pecuniary market and the public economy. This is used as an explanation for the lack of recognition of this economy (Cheal 1988).

A detailed discussion of whether inheritance, giving and receiving used clothes can be considered to be a gift would require an article in itself. This depends on, among other things, how "gift" is understood (Arnould and Rose 2016). Inheritance is in some ways similar to gifts; it is for example common to show gratitude (Visser 2008), but several factors indicate that these are not gifts, at least not as Belk uses the term. One thing is that this is discussed with different terms and called "inheritance", and they talk about "accepting" the clothing, not only receiving it. The latter points to the fact that the giver wishes to get rid of the clothing and save himself or herself from the discomfort of keeping things that are not in use, or throwing away usable clothes (Laitala 2014).

Inherited and new clothes for the children are not used in the same manner in Lise's family. Lise says this about her daughter's equipment for alpine skiing; note that she uses "we" to describe who "owns" these clothes. "We probably use the best trousers for alpine, and then she has about two that are inherited, what is it called, like down-filled pants or thicker pants that she wears to school and for playing. We do not use the best and finest ski pants for school. So, there are about one or two more of that type of thick pants (…) We probably inherit those she uses in school that are maybe a bit more like, worn."(Lise). The inherited pants function here as second grade both in the way Lise talks about them and in the way they are used. She searched for all clothes that are used for the relevant activities, but not all these clothes were retrieved. In addition, she is (or pretends that she is) uncertain of how many pairs there actually are. In other words, she has a set of prioritized clothes that are new and are used where the clothes are important. In addition, she has older clothes that she has less overview over and that are used for multiple purposes. The difference in priorities is also evident in what the different trousers are called. Alpine trousers have a name based on what they are used for, while she is more uncertain, unclear and general when discussing the inherited trousers.

Also, in her own sport wardrobe there is a system where the newest clothes are prioritized, and these are the clothes we got to see. The clothes from last year or previous seasons function as a supplement, and were not found or displayed.

Lise evaluates the garments based on whether they are inherited and how old they are, but also based on garment type. "I pass on the woolen garments and technical underwear as inheritance, and I could have received used long underwear for my kids, but maybe not for myself (laughter). However, for the children I could easily have done so". She accepts most clothing for her children, but "wool socks I'm not so keen on inheriting from other children" (Lise). In other words, children can receive used underwear, while the laughter indicates this is embarrassing if we're talking about adults. Material also plays a role. Underwear in wool and synthetic technical materials is OK to inherit for the children, but not underwear for everyday use (i. e. cotton).

The quantitative material showed that children receive more clothing than adults, and that younger children get more than older children (Laitala and Klepp 2014). In addition, it is less common to inherit clothing that is worn close to the body, such as underwear. Therefore, it is likely that Lise is not the only one who perceives the used clothing for children to be different and simpler than used clothing for adults. This meaning goes beyond the purely practical aspect of young children growing faster and thus constantly having things they cannot use anymore (Klepp and Storm-Mathisen 2005).

Dina tells of a similar hierarchy: "I'm not that fond of inheriting briefs. My daughter has received one pair from her friend that is couple of years older. So, it was like that; ah mmm, please. And then we received them, and we boiled and inspected and worked with them." This implies a thought that used clothing is less "pure" and therefore less suitable for next-to-skin garments. They must undergo a cleansing process with "boiling and inspecting". We have learned from Douglas (1984) that cleanliness is a cultural construction. Others' underwear is culturally unclean. However, if we see the clothes as something material, the old clothes may be cleaner than the new, because a larger proportion of potentially harmful chemicals are washed out (Greenpeace 2012).

Lise also has aesthetic requirements for what she wants her kids to inherit. The aesthetic requirements are not formulated in terms of what she thinks looks nice, but what is currently available to be purchased. "If it is similar to the new products that you can buy in stores today, so, yes please. But it is because I want the kids to have the same as I have. There is no reason why I should have something better than they, I think. I cannot demand that they should love to do things while using poor equipment if I have something really nice." (Lise). It is good that children inherit, but only if they get things that look like new. Such a position makes inheritance obviously more problematic, because that is precisely what inheritance does:

it ensures that garments are used and owned for a long period of time. On the other hand, we must remember that this discussion with Lise concerns clothes for sports. Sportswear has to a large extent developed more rapidly than other clothes, in functional and aesthetic terms. (Klepp 2010).

Lise is the only one who explicitly mentions the changes in sports clothing as a factor that makes inheritance less likely. While discussing the use of the same garment for multiple purposes, she tells of her son and the use of a hooded fleece sweatshirt: "Specifically this one, he uses in school and for downhill skiing and there is a hood on it, and it is a bit cool" (Lise). Outdoor clothing that can also be used daily is positive "at least for kids".

Anne inherits from her mother: "It is a very common practice that mom gives me a lot she thinks that I should inherit." However, there is a "but". "Out of the equipment for trips outdoors it is only the shoes [that she has accepted]. My parents have never had such outdoors… like the tracksuit. I have no plans to inherit it" (Anne). The tracksuit here implies her parents' leisure wear, a type of colorful and soft two-piece outfit with smooth synthetic fabric lined with thin cotton jersey. This type of garment is outdated both functionally and aesthetically.

Inheritance has a dual status in the interviews. "I am very much in favor of inheritance," says Dina. And most of them speak positively about inheritance. Meanwhile, it is also used as a negative term, in the meaning of old, and often referred to in childhood memories. "I had perhaps an inherited anorak and old knickerbockers [baggy knee pants]," says Elin. "Very plain. The clothes were inherited and there could be five successors." Inheritance as something old-fashioned is also evident in Anne's narrative of her mother. "Her generation is a bit concerned that things are inherited and not purchased new all the time."

The supply of used clothing is dependent on whether you have younger or older siblings. "The youngest inherits a lot. If it is possible to use it again, he inherits it." But the older brother also inherits some from the mother, Beatrice, and then passes it on to his little brother. But it does not stop there. Things that are "in good condition are given to my sister's children," Beatrice explains. Thus, the age order among cousins is also crucial for the supply of inherited clothing. As the inheritance goes beyond the circle of family and relatives, the eldest or only child may also inherit. Dina's son, for example, inherits from "a couple of buddies" of hers.

Statistics indicate that sharing of sports equipment is especially common among children, and that among adults it is more common that the younger respondents inherited sporting goods and sportswear than the older respondents. Inheritance of sportswear is not so common among adults as inheriting other types of textiles such as everyday clothes (Laitala and Klepp 2014).We do not have any good statistical material for measuring change over time, but there are no indications that would suggest a decline in the scope, rather the contrary. The increase in the amount of clothing also means that a greater proportion of clothes are used less and thus can be passed on before they wear out (Klepp and Laitala 2016). If the young adults who now share more than the elderly continue this practice even as they grow older, it is possible that the extent of the practice will increase in the future.

Borrowing also takes place outside the household, e. g. when visiting others. Dina does not have clothes for skiing, but this does not prevent her from going skiing. "When we are with mom and dad I just borrow clothes there, I fit into mom's stuff and Daniel can use dad's things," she says. Daniel, her son, does not have the same resistance to using his grandfather's clothes as we heard when it came to using hers. It is not known if this difference is because of the jacket itself, or the fact that it is the one his mother always uses, that it belongs to the mother, or the presence of friends that makes this difference. However, the story shows that borrowing clothes may be perceived differently.

As we have seen so far, most clothes are shared within households and outside the market sphere. Therefore, it is perhaps not surprising that none of the informants partook in the "new" sharing forms, where digital platforms are used for sharing. In this area, the development is exponential, and great changes have occurred during the last two years alone.

Shared use/access, or shared ownership

A division between owner and user, or the person with access, immediately raises the question of what it means to "own". Ownership within a family is not straightforward. Sharing of a variety of consumer goods is completely general and widespread within the family (Belk 2010), and they can be called "commons" (Gudeman 2001). But does this also include clothes? And is it conceivable that within the family there is one owner, and another person who uses the clothes?

The line between borrowing and owning together is very flexible. Dina says this about some jackets: "This is also mine, thus, it is ours." For Dina, the fact that her son has started using her jackets qualifies them to be seen as "ours". We have seen several examples of mothers talking about their children's clothes as if they own them, even when it was clear that only the children used them. Concerning inheriting clothes for the children, Lise says that there is "nothing I could not have accepted". It is thus she, and not the 7 and 9 year old children, who accepts or rejects inherited clothes.

In connection with inheritance of children's clothing, we talked about Anne limiting the oldest child's use of a ski suit. It still needs to be nice when the youngest inherits it. Mothers' management of children's wear in this way supports an understanding of children's clothing being her property, and "inheritance" in this case is not a change of owner, but only change of user. It therefore resembles more a form of a loan than a gift. Historical studies have shown that mothers' provision and control continued for wage-earning adolescents living at home (Rose 2013) and are important for teenagers' clothing habits today (Klepp and Storm-Mathisen 2005). Babies' and toddlers' clothes as mother's "property" can also be understood as an extension of the physical connection involved in pregnancy and breastfeeding. These clothes thus have close ties to what Belk describes as a prototype for sharing: mothering (Belk, 2010).

The close link between mother and children's clothing is also seen in economic expenditures. Children's clothing has been perceived as mothers' "private expenditure" (Rose 2013), and this consumption has been important for women's relationships (Clarke 2000). It is therefore not so unthinkable that mothers "own" the children's clothing. It is likely that in many families it will be impossible for the father or the child to give away, throw away or otherwise make decisions relating to the child's clothing. In that case, one person owning and another person using is more common than we think.

The perceived norm for clothes, shoes and accessories is to have one person as owner and user, which also is quite usual. This form of ownership is considered obvious. For grownups, it is perhaps also the most common, although we do not have much research or knowledge on this. We did not ask our informants if they have clothes that no one else is allowed to use, or that no one else has used or will be using later either. Nevertheless, we may assume that much of the clothing and equipment is used by one person, and within the family which shared the most as well. Daniel, Dina's son, is very interested in football, and Dina wanted to give her son the soccer shoes he wanted. The shoes were expensive, "but he paid a part himself. I got a discount, and a friend of mine contributed, so it was a birthday gift from me, the friend and Daniel himself". It is of course possible that these shoes are going to be used by someone else afterwards, but when we visited the family, Daniel was the proud owner and user of these shoes. It is unlikely that he would lend them or that they would end up forgotten and in the lost and found box at school.

In the discussion about "stealing" we heard about "a heap of gloves". The father and son did not entirely agree on the ownership. While the son Christoffer believes ownership is simple, the father practices shared access. The example shows that within the family it is not certain that all agree on what kind of owner and user relationships apply.

One owner and several users are found both within pecuniary (renting) and private (borrowing, stealing or "stealing") markets. These forms of owner/user relationships create flexibility, because multiple users can alternate and thus utilize the clothes in the best possible way. Renting was not discussed much in the six families, while all three forms of informal sharing were common.

It is possible to wear clothes whose owner is unknown. This is of course less likely to be within families, where the number of possible owners is limited, even though it may not always feel like that when the socks are sorted. However, clothes without a known owner may truly be ownerless. Boxes full of clothes in schools and at public swimming pools are examples of this.

Dina was the only one who said that she wore clothes of unknown ownership. The example cited was Dina's wind-stopper jacket, which she uses for most of her activities, and which we heard mentioned before because her son did not like to borrow it. Dina has lived in housing collectives "my entire life, even after I had children. And often I have been the last to move out, so I've got two jackets on top of the last moving load; okay who's is this? – it is nobody's – OK, Dina you take that. So I have two jackets that are waterproof and windproof. The size would fit an elephant though. But it's okay. They work" (Dina). Jackets are so wide and flexible that they can be used for many different purposes, both summer and winter, and have become Dina's because she is the one who uses and stores them. In other words, the "ownership" is constructed through the use of clothes even though they belonged to someone else to start with.

The sharing's "timing"

The timing of clothing consumption is a complicated interaction between access, use and storage in the wardrobe until disposal, and the circulation of use, laundering and care which also entails storing (Boyd and McConocha 1996; Klepp 2001; Winakor 1969). Our method is not geared towards studying the timing of the consumption of clothing, and the information we have obtained on this is random. We will here pinpoint that timing is vital not just for sharing, but also for effective use of clothing by one user.

One aspect of timing surfaced in our material: Anne says that she has previously stored boxes of clothes in the basement when seasons change. However, "then we just forget them. We cannot be bothered to go down to retrieve winter and summer clothes". Memory is certainly reliant on time, and with a lot of clothing, or long storage periods for other reasons, things will disappear from consciousness. Or as Anne says here, she has clothes she does not remember she owns and that she does not use.

The different sharing forms we have described so far have different timings. The sharing occurs in the form of multiple users using the garment interchangeably, as in the example of Dina's large wind-stopper jacket that also acts as reserve for her son. The sharing also occurs sequentially by one person using the garment first, followed by the second user. And we have seen both examples of the clothing item going back to the first user, as in Lise's use of her husband's clothes when she was pregnant, or when something is passed on to a third person, like Anne's pair of hiking boots, inherited from her mother who had inherited them from a co-worker.

Motivation and consequences

When we consider the informants that owned the least, borrowing, renting and sharing provided a flexibility that was crucial for them to be able to take part in leisure and sports activities. When Dina and her son borrow ski clothing and equipment from her parents, this is nice and fine, and if they had not had anybody to borrow from they would most likely have refrained from going skiing. She cannot afford clothes for "this and that". If her son could not borrow her big wind-stopper jacket when his clothes were wet, he would have had to stay home. Motivation is thus linked to the possibility of taking part in activities.

A similar outcome can be seen in the quantitative material. The results showed that inheritance of clothing and sports equipment had some economic significance, especially in low-income families, and that it was more important for children's wear than clothing for adults (Laitala and Klepp 2014). Belk points out that "Sharing can take place under conditions of excess, but it can also occur under conditions of insufficiency". (Belk 2007, p. 131). But what is the motivation for those who have good incomes? This may have to do with the feeling of doing something "right", a type of economic sense of using the resources in the best possible way. Another thing is that it contributes to greater clothing stock and thus more flexibility, of which we have seen several examples, such as the inherited clothes that make it possible to "save" the finer slalom clothes for use only on the slalom slope. We also believe that the various forms of ownership help to maintain social relationships and contribute to closer ties between those who receive and give, lend and "steal" from each other, even if our material is not good enough to study this in detail.

Garments having several users could have positive environmental consequences if the multiple use contributed to increased utilization. The environmental benefits are retrieved by replacing the purchase of new clothes. Based on the observations of the different families, this is not the case for all sharing. Dina says she buys winter clothes for her son "if he does not inherit some things," but the same does not necessarily apply for more specialized sportswear. Among those who have the least, not being able to inherit would not have contributed to increase in purchases, but that they could not have taken part in the specific activities. The same also applies to those who have the most. Here inheritance contributes to a second set, but does not replace new purchases. But for some, inheritance, loans, sharing etc. have such an effect. If Lise had not had the opportunity to use her husband's skiing outfit when she was pregnant, she would most likely have bought a bigger one. In other words, sharing sports clothes gives some the possibility of taking part in the activities, while for others it contributes to fewer purchases of new clothes or to the use of more clothing. It is likely that the environmental importance will be different not only for families in different economic circumstances, but also for various types of clothing.

Conclusion and further research

We have shown different sharing practices based on a qualitative study on leisure wear, and tied this to statistics about clothing sharing in Norway. Our central argument is that the forms are numerous and have an established place in Norwegian clothing culture. Quantitative data on clothing consumption in Norway shows that sharing is common. Sharing of sports equipment is especially common among children, and among adults the younger respondents more commonly inherited sporting goods and sportswear than the older respondents.

We argue this on the basis of the extent of sharing, and due to the fact that many of these forms have their well-known and commonly used denominations, in addition to the extent that the quantitative material shows. Because many diverse sharing forms of clothes are practiced, we think clothes are suitable for discussing sharing forms and the differences between them. This will demand research into more types of clothing than we have empirical data on. Our material indicates that there are great differences between people of different ages, gender and income, between different types of sharing, and between the types of clothing involved.

There is thus a disparity between the forms of sharing practices and those that are studied. Public awareness of environmental issues caused by clothing has increased in recent years, both as a research topic and in debate. Here especially selling and buying of second-hand clothing has received much attention. This is probably because it most closely resembles buying new clothes. There are only a few previous studies on sharing of clothes outside the pecuniary market and inside the home and with family and friends. In our study, none of the families used digital platforms for sharing. The quantitative material on the topic also showed that sharing is much more common among close relatives than in large impersonal networks. Interest in new forms that involve digital platforms is increasing. However, the biggest problem is the amount of new clothes that are purchased. Instead of searching for new solutions that most resemble existing practices, we should look for solutions where the environmental benefits are greatest, or for solutions that are already established and accepted culturally.

We see a link in our material between divided work and control of the clothes in a household, also those one does not oneself use (Yates 2016). We therefore believe that increased sharing outside of households and through commercial systems may be easiest to achieve where divided work already exists in some form. This might be in institutions such as the health sector, the military and some work-places, where divided work already exists, but could also include settings where sharing has practical advantages, as in kindergartens.

Inheritance, sharing, lending and "stealing" are all forms that help to increase the number of users of clothes. It is striking that the various forms are so widespread and so diverse. We therefore believe that a larger study of the cultural conditions of sharing should be conducted. We need to know more about motivations, limitations and what is needed to increase sharing.

This also involves examining the clothing characteristics that can contribute to more sharing. Our material already showed that gendering of sports gear is working against inheritance within the family, because it reduces the number of possible users. We believe that a systematic review of the relationship between products and their social life could reveal several similar design challenges. Sportswear has undergone a development where a greater degree of specialization is essential (Klepp 2010). For other types of clothing, the development has gone in the opposite direction. There is hardly any occasion in everyday life where it is not possible to wear jeans or a white shirt. Therefore, the different types of clothing present different challenges as to how their potential use time could be maximized. We need studies where different types of clothing and users are involved.

We started the chapter by quoting Belk, who pointed out the need for research that clarifies the boundaries between different forms of sharing. To limit collaborative consumption only to "new" forms, or those that do involve money transactions, is not a good idea. Seen from the individual's perspective, the need for a larger ski suit during pregnancy can be solved either by borrowing it from the husband or by purchasing it new or used. The many varieties may also help to solve the same problem of access to clothing. This does not mean that it is not interesting to study the boundaries between them. Based on our material, there seem to be clear concepts of what "stealing" and "inheritance" are, but the boundaries of ownership are more confusing, so that lending and owning may somehow merge. The relationship between inheritance and gifts, and inheriting clothes as gifts, could also be studied more in order to better understand the mechanisms of sharing.

Sharing within the household, outside the market, appears still to be the most important and also the most understudied form of sharing. For a more systematic mapping of these forms a good place to start might be the study of relationships between access and ownership, and between different temporalities.

References

Aall, C., Klepp, I.G., Støa, E., Engeset, A.B., & Skuland, S. (2011). Leisure and sustainable development in Norway: part of the solution and the problem. *Leisure Studies, 30*(4), 453–476.

Arnould, E.J., & Rose, A.S. (2016). Mutuality Critique and substitute for Belk's "sharing". *Marketing Theory*, 16(1), 75–99.

Belk, R. (2007). Why not share rather than own? *The ANNALS of the American Academy of Political and Social Science*, 611(1), 126–140.

Belk, R. (2010). Sharing. *Journal of Consumer Research*, 36(5), 715–734.

Boyd, T.C., & McConocha, D.M. (1996). Consumer Household Materials and Logistics Management: Inventory Ownership Cycle. *Journal of Consumer Affairs*, 30(1), 218–218.

Brace-Govan, J., & Binay, I. (2010). Consumption of disposed goods for moral identities: a nexus of organization, place, things and consumers. *Journal of Consumer Behaviour*, 9(1), 69–82.

Briel, T.S. (1999). *Loppemarked – mye mer enn man tror... En studie av loppemarkeder i Oslo [The Flea market- more than meets the eye... A study of Flea markets in Oslo]*. Retrieved from http://sifo.no/files/file48431_rapport_nr_4-99_web.pdf

Cassidy, T.D., & Bennett, H.R. (2012). The Rise of Vintage Fashion and the Vintage Consumer. *Fashion Practice: The Journal of Design, Creative Process & the Fashion*, 4(2), 239–262.

Chapman, A. (1980). Barter as a universal mode of exchange. *L'homme*, 33–83.

Cheal, D. (1988). *The Gift Economy*. London: Routledge.

Clarke, A. (2000). Mother Swapping: The trafficking of nearly new children's wear. In P. Jackson, M. Lowe, D. Miller & F. Mort (eds.), *Commercial cultures: economies, practices, spaces* (pp. 85–100). Oxford: Berg.

Corrigan, P. (1989). Gender and the gift: the case of the family clothing economy. *Sociology*, 23(4), 513–534.

Corrigan, P. (2008). *The Dressed Society : Clothing, the Body and Some Meanings of the World*. London: SAGE Publications.

Cruz-Cárdenas, J. (2013). Gender Differences in Motivation and Product Disposal Methods in a High Masculinity Collectivistic Environment. *Revista Brasileira de Marketing*, 12(2), 158–179.

Cruz-Cárdenas, J., González, R., & Gascó, J. (2016). Clothing Disposal System by Gifting Characteristics, Processes, and Interactions. *Clothing and Textiles Research Journal*, 0887302X16675725.

DeLong, M., Heinemann, B., & Reiley, K. (2005). Hooked on Vintage! *Fashion Theory: The Journal of Dress, Body & Culture*, 9(1), 23–42.

Douglas, M. (1984). *Purity and danger: an analysis of the concepts of pollution and taboo*. London: Ark Paperbacks.

Ekström, K.M., Gustafsson, E., Hjelmgren, D., & Salomonson, N. (2012). *Mot en mer hållbar konsumtion: En studie om konsumenters anskaffning och avyttring av kläder [Towards a more sustainable consumption: A study of consum-*

ers' acquisition and disposal of clothing]. Retrieved from http://bada.hb.se/bitstream/2320/10630/1/Vetenskapnr20.pdf

Entwistle, J. (2000). *The fashioned body: fashion, dress, and modern social theory*. Cambridge: Wiley-Blackwell.

Felson, M., & Spaeth, J. L. (1978). Community Structure and Collaborative Consumption: "A Routine Activity Approach". *The American Behavioral Scientist*, 21(4), 614.

Fisher, K., James, K., & Maddox, P. (2011). *Benefits of reuse case study: Clothing*. Retrieved from http://www.wrap.org.uk/sites/files/wrap/Clothing%20reuse_final.pdf

Fiske, A. P. (1991). *Structures of Social Life: The Four Elementary Forms of Human Relations*. New York: Free Press.

Fletcher, K. (2008). *Sustainable fashion & textiles: Design Journeys*. London: Earthscan.

Gracey, F., & Moon, D. (2012). *Valuing Our Clothes: the evidence base*. Retrieved from http://www.wrap.org.uk/sites/files/wrap/10.7.12%20VOC-%20FINAL.pdf

Greenpeace. (2012). *Dirty Laundry: Reloaded*. Retrieved from http://www.greenpeace.org/international/Global/international/publications/toxics/Water%202012/DirtyLaundryReloaded.pdf

Gudeman, S. (2001). *The Anthropology of Economy: Community, Market, and Culture*. Oxford: Blackwell.

Hiller Connell, K. Y. (2009). *Exploration of Second-Hand Apparel Acquisition Behaviors and Barriers*. Paper presented at the 66th Annual Conference of the International Textile and Apparel Association, Bellevue, Washington USA. http://krex.k-state.edu/dspace/bitstream/handle/2097/7995/KYHCITAA20092nd.pdf?sequence=1

Jenss, H. (2005). Sixties dress only! The consumption of the past in a retro scene. In A. Palmer & H. Clark (eds.), *Old clothes, new looks* (pp. 177–195). Oxford: Berg.

Kaiser, S. B. (1997). *The social psychology of clothing. Symbolic appearances in context* (Second edition ed.). NewYork: Fairchild publications.

Klepp, I. G. (2001). *Hvorfor går klær ut av bruk? Avhending sett i forhold til kvinners klesvaner [Why are clothes no longer used? Clothes disposal in relationship to women's clothing habits]*. Retrieved from http://www.sifo.no/files/file48469_rapport2001-03web.pdf

Klepp, I. G. (2006). *Skittentøyets kulturhistorie-hvorfor kvinner vasker klær*. Oslo: Novus forlag.

Klepp, I. G. (2010). Sports and dress. In L. Skov (Ed.), *Berg Encyclopedia of World Dress and Fashion: West Europe* (Vol. 8, pp. 382–386). Oxford: Berg.

Klepp, I. G., & Bjerck, M. (2014). A methodological approach to the materiality of clothing: Wardrobe Studies. *International Journal of Social Research Methodology*, 17(4), 373–386.

Klepp, I. G., & Laitala, K. (2016). *Klesforbruk i Norge*. Retrieved from http://www.sifo.no/files/file80519_fagrapport_nr._2-2016_rapport_klesforbruk.pdf

Klepp, I. G., Laitala, K., Schragger, M., Follér, A., Paulander, E., Tobiasson, T. S., … Kiørboe, N. (2015). *Mapping sustainable textile initiatives and a potential road-*

map for a Nordic actionplan. Retrieved from http://dx.doi.org/10.6027/TN2015-545

Klepp, I. G., & Storm-Mathisen, A. (2005). Reading fashion as age: Teenage girls and grown womens Accounts of Clothing as Body and Social Status. *Fashion Theory: The Journal of Dress, Body and Culture, 9*(3), 323–342.

Laitala, K. (2014). Consumers' clothing disposal behaviour – a synthesis of research results. *International Journal of Consumer Studies, 38*(5), 444–457.

Laitala, K., Boks, C., & Klepp, I. G. (2015). Making Clothing Last: A Design Approach for Reducing the Environmental Impacts. *International Journal of Design, 9*(2), 93–107.

Laitala, K., & Klepp, I. G. (2014). Arv og deling av klær, sko og sportsutstyr. In R. Lavik & E. Borgeraas (eds.), *Forbrukstrender 2014: SIFO survey* (pp. 25–28). Oslo: SIFO.

Laitala, K., & Klepp, I. G. (2017). Clothing Reuse: The Potential in Informal Exchange. *Clothing Cultures, 4*(1), 61–77.

Laitala, K., Klepp, I. G., Morley, N., Meistad, T., Chapman, A., Chen, W., … Austgulen, M. H. (2012). *Potensiale for økt materialgjenvinning av tekstilavfall og andre avfallstyper (Potential for increased material recycling of textile waste and other waste types)*. Retrieved from http://www.sifo.no/files/file78453_fagrapport_2-2012_rev1.pdf

Lien, M. E. (2004). The virtual consumer: Constructions of uncertainty in marketing discourse. In C. Garsten & M. L. d. Montoya (eds.), *Market Matters: Exploring cultural processes in the global marketplace* (pp. 46–69). New York: Palgrave Publishers.

Mauss, M. (1970). *The Gift: Forms and Functions of Exchange in Archaic Societies*. London: Cohen & West.

Morley, N., Bartlett, C., & McGill, I. (2009). *Maximising Reuse and Recycling of UK Clothing and Textiles. Incl. Appendix 1 – Technical Report*. Retrieved from http://randd.defra.gov.uk/Default.aspx?Menu=Menu&Module=More&Location=None&ProjectID=16096

Nordic Council of Ministers. (2015). *Well dressed in a clean environment: Nordic Action plan for sustainable fashion and textiles*. Retrieved from http://norden.diva-portal.org/smash/get/diva2:819423/FULLTEXT01.pdf

Palm, D., Elander, M., Watson, D., Kiørboe, N., Salmenperä, H., Dahlbo, H., . . . Rydberg, T. (2014). *Towards a Nordic textile strategy – Collection, sorting, reuse and recycling of textiles*. Retrieved from http://beta.diva-portal.org/smash/get/diva2:720964/FULLTEXT01.pdf;jsessionid=3cc27b5e683d092c8607776da76e

Parry, J. (1986). The Gift, the Indian Gift and the 'Indian Gift'. *Man, 21*(3), 453–473.

Polanyi, K. (1957). *The Great Transformation*. Boston: Beacon Press.

Price, J. A. (1975). Sharing: The integration of intimate economies. *Anthropologica, 17*(1), 3–27.

Rose, C. (2013). Maternal consumption: A view from the past. *Journal of Consumer Culture, 13*(2), 178–198.

Stephens, S. H. (1985). *Attitudes toward socially responsible consumption: development and validation of a scale and investigation of relationships to clothing acquisition and discard behaviors.* (Doctoral dissertation), Virginia Polytechnic Institute and State University, Blacksburg.

Tinson, J., & Nuttall, P. (2007). Insider trading? Exploring familial intra-generational borrowing and sharing. *The Marketing Review,* 7(2), 185–200.

Ulväng, M. (2012). *Klädekonomi och klädkultur: böndernas kläder i Härjedalen under 1800-talet.* Möklinta: Gidlunds förlag.

Visser, M. (2008). The gift of thanks: The roos, persistence, and paradoxical meanings of a social ritual. Toronto: HarperCollins.

Wahlen, S., & Laamanen, M. (2017). Collaborative Consumption and Sharing Economies. In M. Keller, B. Halkier, T.-A. Wilska & M. Truninger (eds.), *Routledge Handbook on Consumption* (pp. 94–105). Oxford, UK: Routledge.

Widlok, T. (2013). Sharing: Allowing others to take what is valued. *HAU: Journal of Ethnographic Theory,* 3(2), 11–31.

Winakor, G. (1969). The Process of Clothing Consumption. *Journal of Home Economics,* 61(8), 629–634.

Woodburn, J. (1998). 'Sharing Is Not a Form of Exchange': An Analysis of Property-Sharing in Immediate-Return Hunter-Gatherer Societies. In C. M. Hann (Ed.), *Property Relations: Renewing the Anthropological Tradition* (pp. 48–63). Cambridge: Cambridge University Press.

WRAP. (2015). *Resource Revolution: Creating the Future. WRAP's plan | 2015–2020.* Retrieved from http://www.wrap.org.uk/sites/files/wrap/WRAP-Plan-Resource-Revolution-Creating-the-Future.pdf

Yates, L. (2016). Sharing, households and sustainable consumption. *Journal of Consumer Culture,* 1469540516668229.

List of Authors

Davide Arcidiacono

Catholic University "Sacro Cuore" of Milano, Milan, Italy

Davide Arcidiacono is researcher in Economic Sociology at the Catholic University of Milan. He is Member of the Scientific Committees of Co-hub Milano and Sharitaly, the biggest Italian convention on the sharing economy. He has published several papers on critical consumption, such as "Consumer's rationality in a multidisciplinary perspective" (2011), "Consumatori attivi: cittadinanza e partecipazione per una nuova etica economica" (2013).

Marcos Barros

Grenoble École de Management, Grenoble, France

Marcos Barros is an associate professor at Grenoble École de Management (France). His research interests include technology and social media, critical perspectives on identity, resistance, and change, alternative forms of organization, and institutional dynamics. His research articles have been published in Organization Studies, Organization, Organization Science, and Journal of Management Inquiry.

Isabel Cruz

Institute of Sociology at the University of Porto, Porto, Portugal

Isabel Cruz has a degree in Sociology (New University of Lisbon, 1983), a Master in Sociology (New University of Lisbon, 1996), and a Ph. D. in Sociology (University of Porto, 2009). During a Post-Doc phase (University of Porto, 2016) she received a grant from FCT (SFRH/BPD/64192/2009), to conduct research on "Sustainable consumption: between expectations and constraints. The role of political intervention and the new social movements" which was completed in September 2016. She is currently developing the project "Collaborative Consumption and Mobility: Urban Policies and Strategies towards the Low Carbon City".

I. Cruz et al. (Hrsg.), *Contemporary Collaborative Consumption*,
Kritische Verbraucherforschung, https://doi.org/10.1007/978-3-658-21346-6

Rafaela Neiva Ganga
Institute of Cultural Capital (ICC) at the University of Liverpool and Liverpool John Moores University, Liverpool, United Kingdom
Rafaela Ganga is Institute of Cultutal Capital Research Associate since September 2016. Formerly she was a Foundation of Science and Technology Research Fellow at the Institute of Sociology; and Lecturer at the Porto Higher Institute of Social Work. She has a degree in Education and PhD in Sociology from the University of Porto. Currently, she is leading research projects on cultural policy, museum studies and cultural consumption.

Gazi Islam
Grenoble École de Management, Grenoble, France
Gazi Islam is Professor of Business Administration at Grenoble Ecole de Management (France). He completed his Ph. D. in Organizational Behavior at Tulane University, where his research focused on organizational identity, voice, and power relations. He has served as faculty at Insper, Tulane University, and the University of New Orleans. His current research interests include the organizational antecedents and consequences of identity, and the relations between identity, group dynamics and the production of group and organizational cultures.

Ingun Grimstad Klepp
Oslo Metropolitan University, Oslo, Norway
Ingun Grimstad Klepp is research professor at Consumption Research Norway (SIFO). Her research focusses on sustainable textile, clothing, laundry and leisure consumption and has written numerous articles and books on these themes. The relationship between textiles, social and physical characteristics and how these are woven together is at the core of her interest. She loves popular dissemination and new methods.

Ariela Mortara
IULM University of Milan, Milan, Italy
Ariela Mortara is Assistant professor of Sociology of Consumption at IULM University of Milan, Italy. She has a degree in Foreign Languages and Literature (University of Trento), and a Master Degree in Corporate Communication (UPA-Ca' Foscari). Her research topics are consumption and consumers, and the relationship between consumption and new media and corporate communication. Some of her most recent publications are: "Net-Activism on Twitter. McDonald's and Coca Cola @Expo2015" (2017 with S. Fragapane), "The Spread Fashion: an Explorative Research of Italian Fashion Blog" (2017 with G. Roberti)

Mikko Laamanen
Royal Holloway University of London, London, United Kingdom
Mikko Laamanen is a lecturer in marketing at Royal Holloway – University of London. His research considers forms of organisation and collective action around work and consumption with particular focus on alternative forms of economic and social collaboration.

Kirsi Laitala
Oslo Metropolitan University, Oslo, Norway
Kirsi Laitala is a senior researcher at Consumption Research Norway (SIFO). Her research interest lies within sustainable clothing consumption including all the different aspects of use from acquisition to disposal. She uses interdisciplinary research methods that often combine material studies of textiles with consumer studies, and qualitative methods with quantitative surveys.

Gian Paolo Lazzer
Department of Management, Ca' Foscari University of Venice, Venice, Italy
Gian Paolo Lazzer has a Ph. D. in Sociology and Social Research from the University of Verona where he teaches Theory and Techniques of Communication. He is a research fellow at Ca' Foscari University of Venice-Department of Management. His research interests focus on consumption, money (in particular new forms of monies) and organisational culture.

Roberta Paltrinieri
University of Bologna, Bologna, Italy
Roberta Paltrinieri is full Professor at the University of Bologna where she teaches Sociology of cultural and communication processes and Sociology of Consumption. She is the Director of Ces.Co.Com., the Center for Advanced Studies on Consumption and Communication of the Department of Sociology and Business Law (University of Bologna).

Geraldina Roberti
University of L'Aquila, L'Aquila, Italy
Geraldina Roberti has a Ph. D. in Communication Sciences from Sapienza University of Rome. She is Assistant Professor of General Sociology at the University of L'Aquila. Her current research interests include consumption and lifestyles, youth subcultures and the relationship between young people's identity and consumption practices.

Francesca Setiffi
University of Padova, Padova, Italy
Francesca Setiffi is Assistant Professor of Sociology of Culture at the Department of Political Sciences, Law and International Studies, University of Padova. Her main research interests concern: consumers' attitudes and beliefs, material culture and design, and cultures of organisations (private and public). She is Member of the Centre of Research of Consumer Strategies of Italian Families coordinated by S.W.G. (Trieste) and C.R.I.S. (Center of Social Research) at the University of Verona.

Stefano Spillare
University of Bologna, Bologna, Italy
Stefano Spillare has a Ph. D. in Sociology and is research fellow at University of Bologna. He actively collaborates with Ces.Co.Com., the Center for Advanced Studies on Consumption and Communication of the Department of Sociology and Business Law (University of Bologna).

Stefan Wahlen
Wageningen University and Research
Stefan Wahlen is Assistant Professor for "Consumption Governance" at the Sociology of Consumption and Households group of Wageningen University, the Netherlands. Stefan's research is concerned with consumption governance, which is the interplay of politics and policies pertaining to consumption. He is mainly interested in consumption practices of eating and sharing and their political realm. He holds a Ph. D. in Consumer Economics from the University of Helsinki (Finland) and a M. Sc. in food and household studies from the University of Bonn (Germany).

Lightning Source UK Ltd.
Milton Keynes UK
UKHW02f1215290518
323385UK00007B/98/P

9 783658 213459